NAOMI GAEDE-PENNER

Prescription for Adventure

ALASKA BUSH PILOT DOCTOR

The Story of Elmer E. Gaede, M.D.

Alaska Bush Pilot Doctor: The Story of Elmer E. Gaede, M.D.
Published by Prescription for Adventure
Denver, CO

Previous editions 1991, 2004, 2008, 2015

978-0-9637030-2-6
BIOGRAPHY / Doctors

Cover photo from the Dr. Elmer E. Gaede collection: Elmer Gaede against his Piper J-3 airplane Lake Hood, Alaska.

Alaska map by Barbara Spohn-Lillo

Tanana map by Anna Bortel Church

Cover and Interior design by Victoria Wolf, wolfdesignandmarketing.com

Printed in the United States of America.

OTHER BOOKS BY NAOMI GAEDE PENNER

The Bush Doctor's Wife

From Kansas Wheat Fields to Alaska Tundra: a Mennonite Family Finds Home

'A' is for Alaska: Teacher to the Territory

'A' is for Anaktuvuk: Teacher to the Nunamiut Eskimos

To Dad, whose oral tradition of storytelling fueled my written storytelling. All in all, we've recorded some family history, entertained a few readers, and hopefully inspired *all* readers to view life as an adventure — or at least like a story.

CONTENTS

AUTHOR'S NOTES

DURING MY GROWING UP YEARS, supper-time was an amalgamation of my mother's fine cooking and my father's supper table tales. Not knowing when the phone would ring announcing an emergency, Dad always wolfed down his food — he didn't want to miss a single bite of moose roast, streaked with yellow mustard, or enormous Alaska grown cauliflower dripping with cheese sauce. Somehow, between bites, he managed to entertain us children, and anyone else at the table, with his true-life dramas.

As a young doctor, and then as a more experienced physician, he thrived on practicing medicine. He'd share with fascination the story of a patient with symptoms that had at first stumped him, but then he recalled something he'd learned in medical school and the opportunity to put into practice that very instant. He was much like a detective solving a mystery — and needing to share the exhilaration of his breakthrough.

If it wasn't a medical story, he'd squeeze out the maximum adrenaline in describing a wild hunting trip. Of course, he'd weave in flying calamities or "just in the nick of time" recoveries.

The stories in this book didn't come from oral renditions alone, but from the years of folded letters he'd sent home to his parents in California, onion-skin documents used in correspondence with the Public Health Service, and brittle, yellowed newspaper clippings. Ten carousals of Kodak slides added to descriptions.

In 1986, he and I turned one tale into an article, which I submitted to *Alaska Flying Magazine.* "No Ordinary Day" was accepted, then two more. These starters were followed by an article in "We Alaskans" of the Fairbanks News-Miner; after that, the *Christian Medical Society* found a fit in their publication.

I naively believed these articles, plus *just a few more* stories could easily grow into a book. After arduous months of research and rewrites, the first edition of *Prescription for Adventure: Bush Pilot Doctor* rolled off the press in 1991. The book, now titled *Alaska Bush Pilot Doctor,* is in its sixth edition.

INTRODUCTION

MY FATHER HAD A RESTLESS, dare devil streak and needed to live on the edge. Growing up on the flat terrain of Kansas, the "edge" wasn't much in regard to the geography. However, one had to consider there was a finger-nail-gripping edge to the farmers' dependency on the capricious weather. Early in life, Dad didn't know he'd someday be on the edge of mountains, glaciers, air currents, floatplane take-offs, and volcanoes. His life appeared predictable.

Admittedly, competitiveness played a part in his personality. He vied with his slightly older brother, Harold, for first place, whether that was running in track, achieving school grades, riding an unbroken farm horse bareback, jumping on an unsuspecting pig and seeing who could stay on the longest, ice skating on a country pond, catching catfish with his hands, or golfing. When it came to swallowing goldfish, his brother won hands down.

Together they walked the railroad tracks to school, getting a thrill out of placing a penny on the track and then, after the endless succession of steel wheels had clickity-clacked past, picking up the smashed circle, flattened beyond recognition.

Inside the school walls, they were shy boys. The contributing factor was their Russian-Mennonite background where Plautdietsch, or "Low German," was spoken at home; accordingly, they were much more fluent in that language than in English. This set them apart. Their responses in classroom life were guarded.

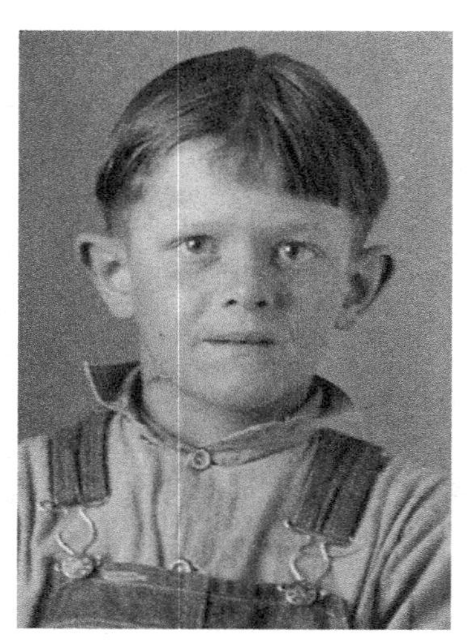

Elmer E. Gaede school picture, Hillsboro, Kansas

For my father's parents, H. A. and Agnes Gaede, living on the edge meant just trying to make ends meet and buying a few groceries. They worked hard at farming, at one point moving from Hillsboro, Kansas to Collinsville, Oklahoma, in search of more fertile land. His mother spent hours in the field, struggling alongside her husband in the blazing sun, begging the hard clods to produce life-giving wheat, oats, corn, and potatoes. "Milk money" from selling cream supplemented their meager income, as did the skimpy sales of eggs and fryers. Too many times she'd return to the small house exhausted, too tired to cook dinner — or not finding enough money in the grocery jar to buy anything.

Whether it was malabsorption or malnutrition during his early years, my father was sickly and missed half a year of school, which had to be repeated and put him behind his classmates. Food *was* scarce. He and Harold fought over soda crackers and considered homemade chocolate pudding an entrée in itself. Surprisingly, as an adult, my father was still very fond of chocolate pudding.

Within a short time, his parents moved back to their roots in Hillsboro, Kansas. Dad was a scrawny kid who grew into what Gaede men like to describe themselves as wiry. Regardless of any physical deficiencies, he had the agility and stamina of a mountain goat, which served him well in the life he later pursued.

The community and culture he grew up in expected that farm boys would transition into farm men and farm girls would marry farm boys. Consistent with this role, he dreamed of owning a 160-acre farm. By his mid-20s, however, this vision had changed to that of being a medical missionary.

But wait, he's better at telling this story than I, so let's listen to what *he* has to say. For now, fasten your seatbelt, pull on a warm coat, grab a ring of bologna, and get ready for Doc Gaede's prescription for adventure!

Naomi

* Soldotna, site of Dr. Gaede's homestead

CHAPTER 1

OPENING THE DOOR TO THE LAST FRONTIER

1955

MY INTRODUCTION TO ALASKA came in July 1955, when my wife, Ruby, our two preschoolers, and I journeyed eight days from the gently rolling prairies of Kansas to the vast but unpopulated frozen wilderness of the Alaska Territory.

"What is a mountain?" my five-year-old daughter, Naomi, had queried repeatedly from her standing-on-her-head position in the backseat of our once-shiny black '47 Fleetline Chevy. No seat belts restrained her restlessness or her questions. Her younger sister, Ruth, sat on a pile of blankets, her curly head bent over her Tiny Tears doll. Naomi's questions were justified. The only *mountains* she knew were the rolling hills of the Kansas University campus at Lawrence, where I had completed medical school.

Ruby, Elmer, Ruth, Naomi at Elmer's graduation from Kansas University Medical School, 1954

As far back as Holland and as recently as my Mennonite great-grandparents who emigrated from Ukraine, farming was in my DNA — as it was in my wife Ruby's. Even so, I hadn't seen a future in tilling the ground — although this innate skill would come in handy later when I homesteaded. Instead, I was fascinated by the medical careers of the Gaede (GAY-dee) doctors preceding my generation. My mother had wanted to become a nurse, but it didn't fit her life roles. When she caught wind of my interest in medicine, she vicariously and determinedly supported every effort. At age 32, I had a new certificate that boldly asserted: "Elmer E. Gaede, MD"

For me, medical work and missionary work seemed to go hand in hand — and Ruby agreed; her heart was bent toward service to others, too. Initially, my objective was to practice in South America, where colonies of Mennonites mixed in with the indigenous people. I corresponded with C.A. DeFehr, at the Canadian Mennonite Brethren Conference in Winnipeg regarding opportunities in South America. The only available position at that time was in Paraguay. The location was suitable, but as a neophyte physician, the responsibilities he listed overwhelmed me. I graciously declined and hoped for a less rigorous position. There wasn't one. Instead, I headed north, to Alaska.

Why did I choose the frozen tundra, rather than the torrid jungles? What influenced my drastic turnabout? An exuberant nurse, who worked with me during my residency at Bethany Hospital in Kansas City, had inundated me with exciting stories of medical possibilities in Alaska. For several years, she had worked at the hospital in Bethel, Alaska, and even now continued to correspond with friends who remained there. Her eyes gleamed every time she told me, "It's a door to opportunity!"

As I investigated these opportunities, I learned that the Bureau of Indian Affairs (BIA) was being replaced by the United States Public Health Service (PHS). Consequently, the Alaskan Native hospitals needed additional medical officers. Furthermore, the Doctor Draft Law would allow my work there to qualify as government service and thus fulfill my military obligations. If that wasn't incentive enough, I'd heard there was good money as a physician with Public Health in Alaska — maybe even $7,000 to $9,000 a year! An unheard-of sum when school teacher's pay was between $3,440 and $3,700. And, I needed every penny to repay school loans.

I requested employment at the Anchorage Native Service (ANS) Hospital. Request granted. One detail remained: I needed licensure to practice medicine in Alaska. I wrote to Sister Charles August, Assistant Administrator at Providence Hospital, who then appealed to W.V. Whitehead M.D., on the Board of Examiners, in Juneau, Alaska.

The letter read like this:

Dear Dr. Whitehead:

Dr. Elmer Gaede of the U.S. Public Health Hospital in Anchorage has applied for temporary licensure to practice medicine in Alaska.

Because of an immediate emergency in our anesthesia department we ask that action on Dr. Gaede's request be taken as rapidly as possible. He will be able to assist us during our emergency if his license is granted soon.

Whatever you can do to make it possible for us to use Dr. Gaede's services will be greatly appreciated.

The door swung open to the Last Frontier.

Edging toward that door, the snail-shaped Chevy crept, ever too slowly for my eagerness. At last, we saw not only *a* mountain, but rows of jagged majestic mountains crowned with glistening ice fields. The steep roads through the northern Rockies tested the flat-land car with hairpin turns. When we arrived at their peaks, we were usually treated to awe-inspiring vistas in the distance and winter snow banks at our feet. Snow caves hollowed out by the summer's warmth begged us to explore their dripping, blue-shadowed caverns. As a kid, we'd played "King of the Mountain" on haystacks. Here, no questions asked, I *was* king of the mountain!

At Dawson Creek, we got acquainted with the wash-boarded and chuck-holed gravel Alaska-Canada Highway, referred to, often with a sense of trepidation, as the "Alcan." The highway extends through the Yukon Territory to Fairbanks, Alaska. This project was no small engineering feat. The highway resolutely trudged over permafrost, muskeg, sphagnum moss, and thousand-foot canyons. Rivers murmured up to the roadsides and then ran back into the forests. Primarily constructed as a military supply route to Alaska for U.S. forces during World War II, the Alaska-Canada Military Highway was opened to the public in 1948, after the Canadians felt assured there were adequate gas stations and lodging for travelers — so they would not get stranded on the infinite unpopulated stretches. Since that time, a steady stream of pilgrims has traveled along the highway to start fresh lives or validate their dreams.

We plowed up the dust-choked highway with "Alaska or Bust" finger-written on the dirty hump of the Chevy. When we crossed the border, we became more specific and wrote "Anchorage or Bust." By the time we neared our destination, Ruby and I had our ears filled with the girls', "Are we there yet?"

Anchorage or Bust!

Later I read of the discussion that arose over this city's name. Woodrow Creek for President Woodrow Wilson was one option. Other people suggested Whitney and Brownsville for two early homesteaders. None of these names actually appeared on the ballot, but instead nine other names were listed, among them Matanuska (Mat-uh-NOO-skuh), Gateway, Ship Creek and Anchorage. "Anchorage" won by an overwhelming majority since the name accurately described the site, which was an anchorage for boats that brought freight to the railroad and the coal mines.

We stopped to camp at Ship Creek, a wonderfully clean, clear river, where we could cleanup, and wash the Chevy, too. We were about to become residents of this city and wanted to be presentable.

When we drove into Anchorage the next evening, we wondered about the strange haze over the city. Then we noticed our black car fading to a powdery gray. We were wrong about the city bringing relief from the dust. Gravel roads ran among dilapidated buildings, sod-roofed log cabins, and modern housing developments. The residue muted the colors of the scattered wild roses and indigo wildflowers. The only roads paved were several downtown avenues and Gambell Street. In spite of this, there were sidewalks along other downtown streets and in some housing developments. We quickly learned, and agreed with the standing joke: "Welcome to Anchorage. Bump. Bump. Bump."

Not knowing where in the city to establish ourselves, we initially checked into the North Star Motel on Gambell Street. Here the clerk made Naomi and Ruth feel at home, gifting them with occasional nickels for a soft spiral ice cream cone from across the street. In the evenings we usually picnicked by the river at Ship Creek flats, at Lake Spenard, or along Turnagain Arm.

Ruby, who was by nature a pioneer, fought kamikaze mosquitoes and cooked hot dogs and canned soups over a campfire. Her farm girl background was rooted in practicality, make-do, and fresh air, and had prepared her for this unexpected twist in life. In the early years of our marriage, we'd lived in plain, difficult, and small situations — such as a damp one-room milk house. I found out right away that her five-foot-two-inches was packed full of fortitude and she had the ability to create home in any circumstance.

Ship Creek, July 1955

While she made supper on the riverbank, I either washed the car — a futile effort — or talked with other campers — a mix of tourists and summer workers, some of whom lived in their cars. Naomi and Ruth clunked rocks tirelessly into the clear rushing water.

"Isn't Alaska great!" exclaimed Naomi one evening. Her bare toes were a cold red and arms were goose-bumped, but her face radiated satisfaction.

"Yes," I answered with a grin. "And I bet it's going to get even better."

Ruby's response was not the same. Even though she *was* of strong inner fiber, she had misgivings, and as we gathered around the fire for supper, she said apprehensively, "Elmer, this is so different from Kansas."

"You'll get used to it," I replied, trying to brush off her concern.

My spunky wife had willingly agreed to move to Alaska, but at this point, she did not share my enthusiasm for where this road to the unknown had brought us. She skillfully stoked the fire without dislodging the frying pan, and then said, "Kansas was open and wide. Here we are trapped. Mountains to the east and water to the south and west."

That was true. These purple mountain majesties and the silver shining sea were not like the vast expanse of Kansas golden fields of grain. This territory attracted some people, while creating uneasiness in others. I hoped Ruby would soon grow more comfortable.

"There is only one real way out to civilization. What would happen if that road were ever cut off?" she rubbed campfire smoke out of her hazel eyes, tried to shake a mosquito out of her shiny brown hair, and sighed in doubt. The still-bright evening sun brought out the natural auburn highlights.

Instead of responding to her hemmed-in feelings, I changed the subject and expounded on how up-to-date Alaska was compared to the way it used to be.

Back in the early 1900s when Alaska needed transportation from interior coal fields to exterior waterways, Alaska's railway system was just beginning. The headquarters was originally planned for Seward, but Anchorage was selected. Word of the new railroad created a kind of construction worker's gold rush. To assist in the construction, nearly 2,000 people rushed up from the States and pitched tents on these same flats. In the summer of 1915, the tent city was ordered to move. At the same time, lots were auctioned off on the south ledge overlooking Cook Inlet. Anchorage was born. We had it good now: a highway through Canada, air travel available, and modern housing. What more could we ask for?

After six weeks, we moved from the motel to 2500 East 16th Street in City View, a new residential area on the outskirts of town. Facing the west, our large living room window opened to a mesmerizing panoramic view. During the summer and fall, three-foot-high magenta fireweed burst out along the road against the backdrop of the Alaskan Range. Within this range rested Mount Susitna, a shapely section appropriately named Sleeping Lady. At times she lay blanketed with puffy white clouds, and during the late summer evenings golden lavender rays covered her.

Later, in winter, we'd find the flower hues across the road exchanged for the rosy glow of frozen sun on subzero snow. Dog sleds would mush across our view, interrupting the still life. Regardless of the season, on a clear day, we could look toward the north and see Alaska's centerpiece: the eternally snow cloaked Mount McKinley, its head held high and poking through a wreath of clouds.

With the primary purpose of being a tuberculosis sanatorium, the Anchorage Alaska Native Service (ANS) Hospital had recently opened on November 29, 1953. The six-story structure on Third Avenue, built in the shape of a cross, dominated the skyline in the eastern part of the city. At that time, treatment for tuberculosis emphasized diet, sunshine, and bed rest. Colors of pink, sky blue, and aqua brighten many walls, and asphalt tiles in red or yellow covered the floors. The solarium on the sixth floor offered a spectacular view of Cook Inlet and the Chugach (CHOO-gatch) Range.

Six weeks after my arrival, the United States Public Health Services replaced the Bureau of Indian Affairs and the 400-bed hospital was opened for Natives with other medical issues besides tuberculosis. Seven of us medical officers had been assigned to develop the various departments.

With all the needs in the hospital, there was a selection of work preferences, and roles were multiple. Anesthesia was not only a critical need, but my choice because I had just finished my internship in this area. In addition, since none of the other officers was interested in obstetrics, I also became Chief of Obstetrics. Chief of Outpatients rounded out my list of duties.

The Native women commonly had 12 to 15 pregnancies, and the Eskimos especially had easy deliveries. As with other American Indians, infant mortality rate among the Alaskan Natives was very high. For Alaskan infants, the rate was three times that for White infants. Some of this was due to pneumonia, early infancy diseases, and accidents. During the past several years, village chiefs had begun sending their wives to the hospital, rather than using the traditional village midwifery to deliver their babies.

My patients, Athabascan (ATH-uh-BAS-kuhn) Indian, Aleut (AL-lee-OOT), and Eskimos provided many opportunities to learn about the life, customs, and challenges of the Native people. Their cultures intrigued me and I felt compelled to share these stories with my family, often around the supper table.

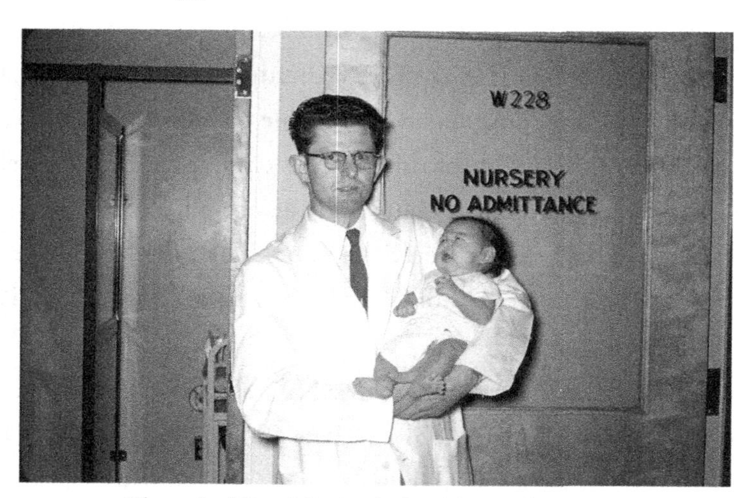

Elmer holding Native baby, November 1956

The standard two-week hospitalization of mother and infant gave the babies an extra boost before reentering village life. While in the hospital, birth defects, especially heart defects, could be detected. During the stay, the mothers were taught post-natal care. When mother and baby did return home, canned milk and vitamins were sent along for the babies.

Shortly after we arrived, and whenever I had a day or two off work, we explored our new environment. Typically, our excursions were short and included summer picnics, and in fall, blueberry picking in rust-colored swamps with white fluffy pompons of wild cotton grass, or red cranberry picking among white birch trees with falling toasted-orange leaves. More accurately, these were lingonberries, but everyone referred to the tiny red berries as cranberries.

In late summer, we drove around the Knik Arm through the Matanuska Valley to Valdez (Val- DEEZ.) At the Knik Arm, an entire field of shooting stars threw out a welcome mat and smiled with little black and white faces, surrounded by fuchsia ruffles. Ruby's eyes lit up at their friendly beauty and of the showy blue lupines with wand-like clusters, which joined the flourishing magenta fireweed along the highway.

We'd heard Paul Bunyan tales of the farming in the Matanuska Valley, which was an experiment emerging out of the Great Depression and carried on under the auspices of the Alaska Rural Rehabilitation Corporation. The concept was noble: to help poor families get a fresh start. In 1935, social workers selected 203 families of hardy Scandinavian descent from Michigan, Minnesota, and Wisconsin to establish farms. The colony struggled with the rigors, unrealistic expectations, and terrain; subsequently, over 50 percent of the original colonist left. Their holdings were taken over by other newcomers. It wasn't until World War II, when Alaska military bases, in particular, Fort Richardson Army Base and Elmendorf Air Force Base, began buying produce that their venture became economically feasible. Eventually, the farms, nurtured with determination and sheltered by the Talkeetna (Tahl-KEET-nuh) Mountains to the north, fertile soil, and Alaska's around-the-clock sunlight, produced abundantly.

We stopped at the State Fair in Palmer. Our mouths dropped open at the sight of 70-pound cabbages, radishes as big as my fist, and cauliflower the size of a basketball. Seeing was believing, and any doubts of the facts were erased.

Our next point of interest was Worthington Glacier, on the Richardson Highway near Thompson Pass. Ruby had planned ahead and packed ingredients for homemade ice cream, along with a freezer. Just listening to the cylindrical metal canister scrape against the glacier ice chunks flooded me with memories of home.

Gaede family: (Bk) Agnes (Ediger), Harold, Henry (H.A.) (Ft) Lillian, Elmer, 1938

In the summer times of my youth, most farm families came into town on Saturday nights to buy groceries. These trips were a respite from the unrelenting responsibilities of life. In the heavy hot air, crickets chirped and other night sounds filled the dark spaces. A mournful train whistle might momentarily take precedence, but was not really heeded since it was so much a part of the background composite. My dad joined the knots of men on the street corners who talked about the weather and speculated about wheat prices. My mother carried a grocery list, scribbled on the back of a letter envelope, and walked down the store aisles. She and other women congregated in front of the flour, sold in cloth sacks, and selected prints that matched a previous purchase or checked to see if a new print was available.

My sister, Lillian followed her around and offered her advice. We young people looked forward to these gatherings — to take a recess from home chores, which often equaled the work of adults, and to see friends and cousins. Our exchanges focused on escapades and adventures, all embellished and meant to impress. Occasionally, my brother Harold and I would use our meager accumulation of coins to buy a bottle of pop or ice cream. I could easily, and with deep satisfaction, spoon away an entire pint myself.

But, most often, we didn't buy ice cream, we cranked it. A large number of our farm community had emigrated, as entire communities, from Ukraine during the late 1800s, so a vast number of us were related. Monthly get-togethers with relatives during the warm months were common. We never lacked for cousins with whom to play — or to fight with for the dasher when it was pulled out of the metal ice cream can.

One time, after a hail storm, my brother, Harold, and I gathered up the golf ball size ice balls and hauled them to our uncle's nearby farm. The destructive ready-made ice that had pelted our crops was turned into something flavorable — ice cream!

Now, here in Alaska we tasted that same flavor of home. Cranking homemade ice cream at glaciers soon became a family expectation, and a comforting tradition that tied our lives together with what had been and what was becoming.

Naomi, Elmer, Ruby, Ruth on Worthington Glacier, September 1956

After indulging in the sweet treat, we walked and slipped on top of the glacier, holding our breath as we jumped across two-foot-wide crevasses and then peering down into their bottomless turquoise depths. I led the way, shadowed by Naomi. Ruby followed with Ruth tightly clasping her hand. Ruth enjoyed the outdoors, but her personality and younger age prevented her from outright exuberance for these explorations. And Ruby's concern for her daughters' safety put her on reserve, too.

To get to our destination of Valdez (Val-DEEZ), we drove through the rugged Alp-like Chugach Mountains. Smudges of clouds grouped together and cut off their peaks, then drifted away, leaving sharp outlines against the royal blue sky. Except for Ruth, who was just as captivated but more subdued, we vocalized our astonishment at the frothy waterfalls, some frolicking gently down the mountainside with angel-hair spray, nearly hidden by the towering dark green spruce. Others demanded to be noticed as they pounded their way into view with a deafening roar. The texture of this country sharply contrasted with our home place roots: the contours of the land and the water features we knew, those of muddy, sluggish creeks and ponds that harbored catfish.

On the edge of town, Ruby exclaimed, "Are those berries?" I down-shifted and pulled the car to the side of the road, against the dense under-brush. We didn't even have to examine them closely. The huge raspberries let themselves be identified easily. Ruby and I opened the car doors and a girl climbed out of the backseat on either side. Ruth hesitated to taste a berry until Naomi poked a quarter-size one in her own mouth, then the two picked and nibbled until their fingers were red-purple and their tongues were stained as well.

After driving through the mountains to get to Valdez, and now seeing how they encircled the bay, we could understand why this area had acquired the nick-name of "Little Switzerland." Its actual name, Valdez, was a mystery until we learned it had been discovered in 1790 by a Spanish explorer and subsequently received its Spanish name. In 1897, thousands of gold seekers bound for the Klondike gold fields had stampeded to this most-northern ice-free port.

Pocketed against this spectacular background and embraced by the nearby Valdez glacier, the town pointed toward Prince William Sound, with houses and log cabins scattered along two gravel streets. A few boats rocked contentedly along the short dock, which jutted out from another street running alongside the waterfront.

No one knew that in 1964 the Great Alaska Earthquake would create horrific tsunamis, which coupled with aftershocks, would trigger a huge submarine slide that would sweep people off the very dock we were standing on, and then undermine the town. Nor did anyone expect that decades later, in 1989, the *Exxon Valdez* oil tanker, would run aground and produce the largest oil spill in U.S. history. All was at peace at this moment with waves lapping gently against the piers and white seagulls bobbing on the scooped surface.

We wandered around the small town and then stopped at Crooked Creek. Here we watched the determined, sickly, spawning salmon, torn by rocks, exhausted by jumping, coming to the end of their journey.

"Why do they do that Daddy," asked Naomi.

I'd never seen such a ghastly, yet absorbing sight and couldn't easily explain it to the girls. Ruby didn't help explain. How could either of us? We were used to catfish that had no plans to go much of anywhere.

"Here we are, at the end of the road with only one way out," said Ruby. Her forehead creased with concern and her mind was *not* on the cycle of a salmon's life. "How do people survive at this end of the world, much less enjoy it. I couldn't live here, pressed in between the water and the mountains." She paused to look up toward the sky. "It's like the mountains lean over on us."

I cleared my throat and started to clarify how in comparison to Valdez, Anchorage *was* civilized, accessible, and spacious. The glum expression on her face and the slump of her shoulders changed my mind, and my voice trailed off. I hoped she'd eventually relax and become as captivated by this frontier, as I was.

Summer ended abruptly. Frost in August. Snow in September. The smiling face of the "Land of the Midnight Sun" turned its cold back on us and everything else in the Territory. In the early years, these first snows, inching down the mountains, were appropriately coined "Termination Dust." When the only transportation between Alaska and the United States was by ship, this white dust, coming to about 3000-feet, signaled the summer workers to head south before freeze-up. Already in October, we were thoroughly immersed in an Alaska winter.

In December 1956, our son, Mark, was born in the Anchorage ANS Hospital. Ruby shared her room with a chief's wife from Savoonga (Suh-VOON-guh) on St. Lawrence Island. "You should see the contrast between your little brother and the Native babies," I laughingly told Naomi and Ruth. "The Native babies are long and thin, and have lots of thick black hair. Mark is chubby, pale white — and bald." To illustrate the point further, I lined up all the babies in the nursery, including Mark, and took a picture.

The hospital served the Native people with medical care and also as a gathering place. With so many Alaskans contracting tuberculosis, and subject to lengthy stays, family members and friends would come from the villages to visit them at the hospital. Also, when Natives flew into Anchorage for other matters, or just lived in the area, they knew they could find recognizable faces at the hospital. I didn't really need to go further to interact with the Native people. When I walked down the hall, I might see a row of potty-chairs with dark-skinned toddlers or a grandma leaning against a wall stitching beads onto a piece of moose skin.

In one instance, the Savoonga chief himself was at the hospital. The Saint Lawrence Island Yupik Eskimos were well known for their finely detailed ivory carving. In exchange for my delivery of a healthy son, the chief, who did not speak English, gratefully opened his sealskin bag and allowed me to choose gifts. I deliberated. The decision was not easy. Finally, I settled on an ivory owl, an ivory bracelet for Ruby, and a fertility stick — a long piece of ivory, etched with Alaskan scenes. These I added to my small collection of ivory salt and pepper shakers, soft fur and skin Eskimo yoyos, and miniature skin kayaks. I was touched by his openhanded kindness.

In February, I wrote home to my uncle in California,

> *We are having about twice as bad weather as we get in Kansas. Yesterday and today we have winds up to 55 mph, and the streets are drifting closed every fifteen to thirty minutes. I suppose the mayor will declare an emergency again and close business tomorrow.*

That blizzard shrieked around us, pounding the doors and windows, not stopping until snow piled up to the roof on the south side of our house. Outside, road graders attempted to clear the streets, and snow blowers swallowed up the ubiquitous white stuff, spewing it out in soft, puffy ribbons.

I wasn't one to stand around and observe from a distance or behind a window, so I bundled up the girls in their snowsuits and we ventured out to nature's winter amusement park, nearly in our front yard. Across the road was an ice-skating area and I put my hockey skates to good use. A half block west, the 30- to 40-foot high gravel pit hills provided great sledding and tobogganing fun. Ruby relished the outdoors and would have joined in, but Mark was too young to be exposed to the deep chill. Someday he'd grow into an Alaskan man, but for now, he was a fussy little guy acting his age.

The Anchorage Fur Rendezvous, held in February, provided more entertainment. Originally a celebration when trappers came to sell their winter's cache of furs, this annual, ten-day cabin fever antidote attracted crowds of Natives and Whites. The hustle and bustle of dogsled races, dog-pull contests, snowshoe races, and fur auctions nearly shut down Fifth Avenue. In one of the open lots there was a platform with hundreds of raw furs, sectioned off for red fox, white fox, mink, beaver, muskrat, lynx, and wolverine.

I thought of my own hunting "trophies." During the dust storm days in Kansas, rodents would destroy our crops. Besides paying up to $2.50 for skunks, the government paid ten cents apiece for squirrel heads and jackrabbit ears. When Harold and I each turned 12, our father allowed us to buy a rifle. We strung our catch on a wire and took them to the county court house in Marion. We felt rich — and we could buy more shotgun shells to increase our pocket change. Dad paid Lillian, our younger sister, ten cents a can for potato bugs that plagued our large potato field. Our nearly two dozen white cats couldn't keep up with the mice so he paid a penny a piece for those critters.

The hides here in Anchorage generated more cash. At the first of the Rendezvous, I bid on the red fox and got two for $5.00 each. The next day, some of the same quality of fox went up to $20.00 each. I was told that I did well to bid early since the furs usually sell low the first few days before the buying interest is up. Later, when the buying fever was aroused, the prices would go up.

Ruth and Naomi with Fur Rendezvous ivory, fox skins, and mukluks, February 1956

When the furs were brought in from the cold and into a warm room, the odor went up too. Some of the furs came from villages where they had been tanned in barrels of human urine. The aroma reminded me of my boyhood hunting. Harold and I would take our catches in only once a month, which meant they hung in a barn-side shed and ripened during that time. I learned that in fur selection, one needs to use both eyes and nose.

I took in as much as I could, but there was so much more for me to witness and investigate. I asked questions and read books. One of the emergency room nurses joked that I might get "Frontier Fever." I raised my eyebrows in question.

"Oh, the prognosis isn't good and some individuals have had to be quarantined in Alaska for the rest of their lives," she said with a laugh.

I had a funny feeling that I'd contracted that — and that I was only in the initial stages. Who knew what full-blown symptoms would be like? I was betting I'd find out soon.

CHAPTER 2

TUNDRA TAXI

February 1956

CRADLING THE NEWBORN BABY in the crook of my arm and carrying a diaper bag, I climbed the steps into the DC-3 and looked around for a seat. No seat assignments. No flight attendants. Besides the flight crew, myself, and the two-week-old baby, I counted five Native men. Wood crates, machinery parts, thick ropes, construction supplies, and other cargo had crowded ahead of us and now occupied seats and floor space. The air was thick with the smell of oily tarpaulins.

"Be glad this plane has a few seats in it," said the pilot, who filled up the cockpit door. "Try to find one with seat belts."

I exchanged greetings with the other passengers. "This is Andrew," I said, gently patting my layered bundle. Their silent, friendly smiles were framed by thick fur-ruffed parkas. Our small group checked seats for seat belts. I managed to locate not only a seat with a belt, but a seat with a window view.

This was my first experience heading into the Interior. Now I was going to see the *real* Alaska. I wasn't sure what to expect, but I didn't want to miss anything. Maybe one of the men could tell me about the Inupiat

(In-OOP-pee-at) Eskimo village of Noatak (NO-uh- tack) — where I was headed. I glanced around. Unfortunately, no one sat near me. I settled in and buckled up. The giant propellers reluctantly ground to a start in the minus 40° Fairbank's air. The baby, undisturbed by the commotion and the cold, continued to sleep in my arms.

This was the second leg of our trip. Several hours earlier, at 9:30 a.m., and just as the February sun was weakly dawning, my small companion and I had boarded a DC-4 and set out on our 750-mile journey, which zigzagged from Anchorage International Airport, north to Fairbanks, west to Kotzebue (KOT- sa- bue), and then to the baby's home in the village of Noatak.

Elmer and baby Andrew ready for trip to Noatak, October 1956

Getting anywhere in winter took a long time, and in 1956, the flights were few and far between; people agreed to fly in anything. Apparently, when heading into the Interior there was no distinction between cargo and passenger planes. Nonetheless, things weren't as grim as in the earlier years, such as in 1942, when the flight between Fairbanks and Nome took six hours with a rest stop about halfway at Galena.

Douglas DC-3s, like the one I was in, first flew in 1935. It was a favorite of pilots both in and out of the military for its relatively large payload, short airstrip requirements, and ruggedness. During the 1950s, the DC-3 and the DC-4 were the most commonly used airplanes by airlines in Alaska. Airlines flying in Alaska (some only regionally) were Northwest Airlines, Pan America World, Northern Consolidated, Reeves, Alaska Airlines, and Wien Alaska Airlines.

I was traveling with the baby because his mother had tuberculosis and needed to remain in the Anchorage ANS Hospital for an indefinite period of time. It was not uncommon for a pregnant Native woman to come to the hospital for delivery, be diagnosed as having tuberculosis, and end up staying in the hospital while their babies went home to be with the other family members. In many cases, family or friends would come get the baby. When this was not possible, one of the hospital medical personnel served as baby courier.

I chuckled to myself. When I'd taken the job at the ANS Hospital, I didn't know "baby escort service" was written between the lines of my contract. Actually, I'd volunteered for this venture and had impatiently waited for the chance.

Andrew whimpered as the plane lurched stiffly off the airstrip. I pulled back the blue flannel blanket, which protected his face from the raw air. Thick black hair crowded around his face. His dark almond eyes blinked and his tiny mouth puckered into a cry. As a father of three young children, I was familiar with the needs of babies, and when the infant continued to complain, I suspected a wet diaper. My breath puffed into a vapory cloud in the frigid cabin, and I decided for the baby's benefit to wait until the temperature warmed to expose him to the cold. Thirty minutes later, I still shivered in the frosty cabin, and the baby cried unrelentingly.

I made my way up to the cockpit and inquired, "Uh, can we expect cabin heat soon?"

Without taking his eyes off the instrument panel, the pilot answered matter-of-factly, "Seems there's a problem with heat out there." The bulky man had his parka hood tossed back and appeared comfortable enough.

I stood for a moment in the warmth of the doorway, waiting to hear what he planned to do about the problem. He turned his head and said over his shoulder, "Hope you have enough clothes on."

Wishing for better accommodations, but realizing I was on my own, I returned to my seat. I would have to make this diaper change as quickly as possible. "There, there," I said in my most soothing voice. "I know this isn't fun. I'm hurrying as fast as I can."

The infant cries intensified. This remedy hadn't solved what I'd hoped for. I dug around for his bottle. I gave the liquid formula the milk-drop-on-wrist test. He was going to be angrier yet with the Yukon-cold liquid. The cockpit had heat and I wondered if we could at least get the chill out of this milk by placing it beside the heater. Once again, I walked up the aisle and apologetically explained my plight to the crew. Perhaps the baby's cries would annoy someone into action. This time, there was more flight hospitality. Within 20 minutes the copilot, a young man with an army-green stocking cap, and the start of an early five o'clock shadow, energetically brought back a hot meal for the baby — something the rest of us could have used by this time. The baby sucked hard and then slept for the rest of the trip.

I scraped the reappearing frost off my window and then pulled a map out of my pocket to try and identify landmarks. From this aerial perspective, I could understand why the Territory had been misunderstood as a refrigerated wasteland. There wasn't much to see — especially in winter with the perpetual snow and small windows of dusky-light. If that wasn't daunting enough, the powerful cold could plunge some places to record-holding minus 80°, which didn't exactly draw anyone outdoors.

I located Kotzebue on my map and finger-traced a line from Fairbanks, trying to determine our location. I'd read about Kotzebue and its population of over 80 percent Inupiat Eskimo. Kititagruk, as it was called in the early years, was accidentally discovered in 1816 by a Polish explorer, Otto von Kotzebue, who was in the service of Russia and looking for the Northwest Passage. This information had historical value, but what interested me more was that Kotzebue was known as the Polar Bear capital of the world. Wouldn't that be exciting to shoot one of those arctic kings? My mind wandered around in possibilities.

After two hours, the plane circled a blurry site which I assumed was Kotzebue, and then landed a mile south of the village. Holding the baby tightly, I waited for the plane steps to be let down. Then, peering into the 2 p.m. dusk, I cautiously stepped out into the arctic air.

"How do we get to the village?" I started to ask one of the men, but the fierce wind with a chill factor of more than minus 60° rudely whipped my wolf parka ruff into my face and snatched away my words, leaving my frozen lips without any desire to repeat the question. So, this was the Interior. Formidable. Instead of a warm handshake of welcome, the arctic wind grabbed me, and pushed through my heavy army parka and wool army pants, past my insulated underwear, and straight to my bones. All the while the baby slept. "He had better get used to this," I mumbled. "This is home."

Ahead of me, the Eskimo men soundlessly and purposefully padded across the blowing snow in their mukluked feet. Where are they going? I wondered. My "bunny boots" crunched and slid along after the group. The large Army issue boots with white felt exteriors and rigid, slick rubber soles were intended to keep a soldier's feet warm for four hours at minus 50°. I'd soon find out if that was true. I strained my eyes and expected an airport terminal building to emerge from the shadows. From there, I'd need to find transportation to the satellite ANS Hospital in the village. I shivered and I wondered how far we could walk in this ice box. Yet, there didn't appear to be any options other than to follow the men with their parka ruffs pulled tightly around their bent heads.

As I tagged along, a small rectangular building appeared. I stepped inside. Wooden benches extended along two walls. Copying the others, I sat down. They seemed to be waiting for something — waiting with assurance and anticipation that something was going to happen.

"We go to village soon," one man informed me.

At least the bodies in the approximately eight-by-twelve shed should soon make some collective heat, I reasoned.

After plenty of time to let the cold work itself deeper, a chugging sound distinguished itself in the howling wind. Was it a diesel caterpillar tractor? The laboring sound continued, growing louder. The shed trembled. Was it going to hit us? Maybe whatever it was couldn't see us in the blowing snow. I glanced at the others. They seemed unalarmed, so I tried to adopt their casual spirit and act calm. The shed jolted. I jumped up. This did me no good since the one small window was near the eight-foot ceiling and too high to see out. The Natives looked at me and then glanced at one another. Andrew cried out. The shed shuddered again.

Finally, one man offered, "Cat hooks to shed. Shed has runners — like sled. Now we go to village."

I wasn't sure what a "cat" meant in this situation, but over time, I learned that Alaskans call any large bulldozer type of equipment a Cat.

Sure enough, the shed jerked, and knocked me off balance. In my haste to seek shelter from the cold, I hadn't noticed the runners on this tundra taxi.

DC-3 and "tundra taxi" pulled by a Cat, January 1956

As we crawled along at about five mph the rumble of the Cat put Andrew back to sleep, all the while I wondered how I should inform the driver of my destination. As it turned out, even without a pull cord, the Cat driver seemed to know where his passengers needed to get off. After several stops, all motion ceased and everything was silent.

The fur-hooded driver poked his head in the door, "Everyone out. This is the end of the route."

Right outside the door was the rambling hospital. What better service could I have asked for? I debarked, but it wasn't until I was inside the comforting warmth that I realized I had never paid or tipped my tundra taxi driver.

Kotzebue Public Health Hospital, January 1956

A congenial Native woman greeted me at the front desk. I introduced myself and immediately two nurses appeared.

"We've been expecting you — and the baby," said the one with a name tag reading "Mildred." Her nurses 'cap was askew on a tumble of thick blonde curls, which she'd tried to pin behind her ears. She took Andrew out of my arms and made soft baby-talk to the child who squirmed without making a sound.

"Yes, his father and the rest of the relatives can't wait to see him," said her companion, who was likewise in her twenties. "We'll notify the village right away." She had efficiency written on her deliberate motions and quick steps.

I wondered what had drawn these two young women to the far north with its harsh environment. I doubted it was gold, the railroad, agriculture, or mining. However, personal adventure or gain weren't the only motivations for Alaska pioneers. Healthcare, education, and other human services persuaded people to come northward, too. Mere dreams of pleasure would have sent idealists racing back to warmer, brighter climes by mid-October.

Side-by-side Mildred and her co-worker strode purposefully down the hall — with Andrew. My arms felt empty without my little companion. Together we had braved wet diapers, cold milk, and the harrowing experience of a tundra taxi.

Reaching out and patting my arm, the capable desk clerk, said, "He's in good hands. We'll arrange for a charter to fly you to Noatak tomorrow morning."

Our camaraderie had not yet ended.

As I waited in the lounge for Dr. Rabeau, director of the 55-bed hospital, I looked out the window. Nearby, several sled dogs trotted in from the Kotzebue Sound. They were carrying what appeared to be small logs of firewood. How peculiar. This was above the Arctic Circle where I'd heard that only tundra thrives and stunted bushes survive. Where were they finding this wood?

Mildred returned, assuring me that Andrew was fine and informing me that Dr. Rabeau would soon arrive.

"Where are the Natives getting the wood?" I asked, pointing out the window.

"That's not wood, Dr. Gaede," she chuckled. "Those are frozen sheefish that are caught through the ice. They weigh from 5- to 50-pounds and make excellent eating."

Just then, the heavy-set, white-jacketed Dr. Rabeau entered. "Hello, Elmer," he boomed. "I imagine you've had quite a day. Won't you join me for dinner and then I'll give you a tour of this place."

Indeed, it had been a long day, nevertheless, the hot meal restoked my energy. "Tell me about Noatak."

Dr. Rabeau explained in his gruff, yet not unfriendly voice, "It's about 50 miles north of here — about 25 miles inland from the Bering Strait, north of the Arctic Circle, the only settlement on the Noatak River — about 40 or 50 people. Used to be a mining and fishing camp."

Here I was, getting deeper and deeper into Alaska.

The following morning, ice fog hung like a sheet over Kotzebue. Frost formed furry sleeves over the skinny arms of bushes and electric lines, etched intricate designs on the window glass, and rimmed the edges of all buildings. Even though the wind, which had chased snow around the village during the night, had stopped its taunting, the foggy cold stubbornly refused to leave.

How do people survive in this uncompromising climate? I wondered as I sipped hot tea in the hospital cafeteria and waited for information about the last leg of my journey. *They're most definitely a resilient and resourceful lot.*

Shortly after noon, the charter pilot sent word that he thought he could make it to Noatak. Instead of taking off from the airstrip where we had landed, he had his plane tied down on the shore ice, a few blocks from the hospital.

A slightly built Eskimo in his thirties cordially greeted us at the airstrip. The hospital personnel had spoken highly of Tommy Richards and I looked forward to flying with him. I hadn't mentioned to anyone, and for sure not Ruby that I was considering taking lessons myself. With this new idea circulating in my mind, whenever I encountered pilots, I plied them with questions: What's the best plane to land in the Bush? What are "tail-draggers"?

Pilots never tire of plane talk, so my questions were usually answered with stories. The stories heightened my resolve to fly and the information gave me the facts. I learned that the best choice for a plane was one with a back wheel — a tail dragger (as contrasted with a nose wheel) which made it easier to land, handle more abuse, and provide more stability on rough or uncertain terrain — such as river ice, sandbars, and uneven terrain.

Tommy's plane was a tail-dragging Pacer. The early Piper-Pacers had a wheel on the tail, whereas later Pacers had a nose wheel — and were called Tri-Pacers. Tri-Pacers can be modified into tail-draggers. I didn't know these details at the time. I just felt confident that I'd be in good hands for this last jaunt.

Andrew and I were the only passengers, so I wasted no time in climbing inside. I was with a *real* bush pilot, and here was another chance to find out more about Alaska! Tommy kept his comments short as he followed the Noatak River, scrutinizing the monochromatic treeless terrain below which blended into the dull sky above. The village was the only permanent settlement on the nearly 400-mile-long river, but in winter, there was no contrasting path of water to guide us — everything had the same bleakness.

After a while, we circled and buzzed what seemed to be just another group of scattered dark spots pressed into the constantly blowing snow. This sparse conglomeration turned out to be Noatak. At the sound of the plane, the villagers scrambled out of their snow-buried sod houses like ants from a disturbed anthill. I could see them race toward a smooth area, which appeared to be the designated airstrip. Taking the cue, Tommy made another pass over that area and came down for a landing. The skis slid smoothly on the snow.

Andrew's father was easy to spot. The short, thickset man had a big grin across his broad face and his arms stretched out toward the baby. I handed off the tiny bundle. The baby was passed around with oohs, ahhs, and laughter, and then Andrew's father and several other villagers asked about the baby's mother. The questions broadened to other news outside the village and Tommy answered those. He seemed to know everyone and obviously everyone liked him.

After a short time, Tommy interrupted the joyous reunion, "Doctor, we've got to get back. There's not much daylight left and if the fog rolls back in we'll never find home. Besides, we don't dare let this engine cool down or we won't get it started."

Andrew's father shook my hand repeatedly. I patted Andrew and swung up beside Tommy. Mission accomplished. My pilot yelled "clear prop" and started the engine. The villagers moved away, smiling, and waving moose-skin-mittened hands. I'd only had a few minutes with those people, but it was as if we were instant friends. I hated to leave.

No surprise to me, Tommy adroitly found his way back to Kotzebue. Sure enough. He *was* one of those indomitable bush pilots.

The trip back to Anchorage was eventful in that there was cabin heat. I was thankful for simple things, which had become big things.

Ruby and the children met me at the Merrill Field Airport. Ruby handed off a wiggling Mark and Naomi bombarded me with questions. "Did you see a bear? Did you sleep in an igloo? What did you eat? Did they give you gifts?" My oldest always had so many questions, and once blurted out, "I'm just so full of words!" Ruth wrapped an arm around my leg.

Ruby wasn't as interested in the sights or process, she wanted to know how the baby fared.

"No problem," I answered briefly. I couldn't wait to tell them how one tiny life had opened up a new world for me. I'd do that later, over supper. For the moment, I said, "You wouldn't believe the complimentary transportation from the airport!"

Before we crossed the street to the parking lot, a taxi pulled up and a woman with a baby cradled in her arms climbed in. Smoothly, the taxi pulled away from the curb and drove out of sight in the early afternoon dusk. What a contrast to my experiences the day before. This trip to the Interior fueled my Alaskan Fever. I had to find a way to get back.

CHAPTER 3

FLYING OR BUST!

Spring 1956

"HEY, DOC, YOU'D LOVE FLYING," said Wally Zimmer, an enthusiastic friend, as we drove away from Lake Hood where his four-place Stinson on floats was anchored to barrel tie-downs.

I was a frequent admirer of the planes that lined the lake. In fact, at this time, the great colorful brood represented 20 percent of the nation's floatplanes, which meant Lake Hood harbored more floatplanes than any other spot in the world. And no wonder, since Alaskans flew approximately 30 times more per capita than the Lower 48 residents.

After less than a year in this Territory with its oftentimes insurmountable terrain, I could see why. Airplanes meant much more than optional or exotic transportation. Vital supplies were carried via air freight. Emergency aid, more often than not, had to be flown in. The Lower 48 states depended on its web of roads and coast-to-coast railway system, while Alaska looked to the skies.

"You could really see Alaska if you'd learn to fly!" Wally continued; his voice grew louder so as to be heard above the clattering gravel on the narrow two-lane road. I agreed. Just living in Anchorage was not enough. Since the main roads out of Anchorage, other than the Alcan we'd found our way up on, either dead-ended in Homer on the Kenai Peninsula, or in Fairbanks, flying certainly played a major part in seeing the majority of the state.

"You know, Doc, a Piper J-3 would be perfect for you! The two-seater is economical, a good plane to train on, you wouldn't have any trouble selling it — and a good resale price, too. Even if you'd decide to keep it, it would be a great plane for hunting because of its slow speed." His sales pitch was convincing.

This late autumn conversation echoed in my mind when I returned from Noatak in the middle of a history-making intense winter. I thought of my flight with Tommy and remembered the feeling of independence and the sense of defiance of man against nature.

Yes, it did seem like a sensible thing to do: buy a floatplane and learn to fly. "Alaska or Bust" had become "Flying or Bust." But no matter my fervor, Lake Hood was a very large frozen pond at this moment and no floatplanes would be maneuvering about for quite some months.

The winter-in-waiting was not wasted. I worked on two projects. First, I read everything I could about Alaskan aviation. I found out that the first flight in Alaska took place on July 4, 1913, when The Aerial Circus, with James Martin, traveled to the tiny log-cabin settlement of Fairbanks. A crowd watched the dusty takeoff of the biplane from the ballpark that stamped a benchmark in Alaskan history.

Ten years later, Ben Eielson, a quiet North Dakota school teacher, not only took off from the same ballpark, but broke a trail across the immeasurable skies of the untamed land, penetrating the air with the never-heard-before drone of an aircraft engine. As the first pilot to cross the top of the world, he established airmail service in Alaska and opened the gates of the remote wilderness to commercial and passenger aviation.

Exciting stuff. I wished I could have been there. I just hoped there might be a few adventures left in the sky for me — whenever I got there.

My second project was to persuade my earth-loving wife that I should fly. I deliberately used Ruby's feelings of being locked in. I pointed out that the plane offered wings of freedom. "Honey, you wouldn't be cut-off by dead-end roads, dark seashores and towering mountains."

She wasn't easy to convince, even though she realized Alaskans accepted air travel as a routine means of transportation. I most definitely did not mention the tundra decorated with planes or the tales of disappearances in Cook Inlet and mountain alleyways — she was well aware of the risks. There was no easy win in this discussion. She countered with other issues, and granted, I couldn't refute her logic about first paying off medical school loans.

I never did persuade her, but in the spring of 1956, I bought a 1947 red, green, and silver Piper J-3 with a 75 hp engine and a controllable pitch wooden prop. The Christmas-tree-colored plane cost $1,500 with wheels and skis and an additional $1,000 for floats.

Finding the right place to take flying lessons was not as easy as I'd anticipated. At the first place I inquired, I was told I would have to learn on wheels before I could tackle floats. At the next place, Barton Air Service, I asked again if they taught initial flying on floats.

"Certainly," the young attendant answered. I signed up. A man in strapped down hip boots walked through the door and glanced down at my paperwork on the counter. His khaki jacket had a tag with "Barton Air Service Instructor" on a front pocket.

"Bring your plane to our dock at 5 p.m. tomorrow and you'll have your first lesson," the middle-aged, thick-shouldered man told me.

Hesitantly I stammered, "I don't know anything about airplanes — I don't even know how to start mine, let alone taxi it to your dock."

At that point, no one could have told me that someday I'd be taking off and landing on rivers, mud flats, rock-littered beaches, or knobby mountaintops.

For my first lesson, I walked along the docks on the shoreline and brought the instructor over to my plane. We began with the basics: starting the plane.

During the following weeks, I diligently studied my flight instruction book and nearly every evening headed to Lake Hood. I'd splash through the water with my hip boots, hop onto a float, pump out the floats, and check for leaks. Satisfied with the results, I'd hand-prop the plane and taxi over to my instructor. Starting the plane and taxiing were easy enough, but braking for the dock was not. Fortunately for everyone and everything in sight, after some very close calls of nearly smashing into the dock, I learned to plan ahead and cut the throttle in time.

Elmer at Lake Hood learning to fly his J-3, September 1956

I had to give Ruby credit. Even though she didn't endorse my new sport, she and the children often accompanied me to the lake. At the young age of eight months, Mark was intensely fascinated by airplanes. His curious wiggling subsided when he watched the planes buzz on take offs and splash when landing. Later, after I achieved my license and he'd fly with me, he'd sit on a rolled up sleeping bag as quiet as a mouse, taking in everything, from the control panel to the cloud formations, to the landscape below.

When I was a kid, a tractor was natural and comfortable to me. There was no distinction between "before I drove a tractor" and "after I learned to drive a tractor." For Mark, this would be said about airplanes

On warm evenings or *hot* days of 60°, Ruth and Naomi took their shoes off and waded in the clear water. They were 16 months apart in age and close companions, yet each had their own personality. Ruth was a nurturer and played second-mother to Mark, doting on his every need. Naomi was a leader; although when she took the lead, Ruth wouldn't be far behind her. Here at the lake, their toes pressed into the sand as they pulled up water weeds that resembled bamboo. "Fishing pole weeds" they called the 12-16-inch stalks that easily pulled apart into shorter segments.

My time spent at Lake Hood was not as tranquil as theirs. One evening, as my instructor and I taxied from Lake Spenard toward the channel to Lake Hood, I saw a pair of floats bobbing on the waves at one end of the lake. "What are those floats doing over there?" I asked. "And why are they upside down?"

"There's a plane hanging under the water beneath those floats," my instructor explained calmly. "It belongs to a student of mine. Apparently, she relaxed the stick upon landing and the plane flipped."

"Did she get out?"

"Oh yes, and I think she learned a lesson she'll never forget."

The chilling scene was still vivid in my mind as we docked at Barton's Air Service, and my instructor climbed out of the plane. Before bouncing off the floats and onto the dock, as he usually did, leaving me to taxi to my tie downs, he casually said, "I'd like you to take the plane up by yourself."

Solo? No! Not today. I silently shouted, remembering the plane beneath the water and looking up at a gray, formidable sky.

Sensing my reluctance, he assured me, "You'll do fine."

And then he turned and walked away.

Slowly, I taxied back through the channel to Lake Spenard for take-off. Even though I knew the seat behind me where my instructor sat was empty, I found myself talking out loud as I went down my checklist: fuel, oil pressure, magnetos. The instructions were clearly fixed in my mind.

Takeoff was not the difficult part, although a floatplane takeoff requires a different technique than a wheel takeoff. I gave the plane full throttle and pulled back slightly on the stick until after a few seconds of acceleration, the floats partially lifted out of the water. I could feel them teetering on the surface and eased forward on the stick, allowing the floats to plane out on top of the water like a water ski. Because of their hydrodynamic design, there was now the least amount of drag in the water, and the plane could more easily lift off that surface. This "getting on the step" felt familiar.

I didn't like the dark sky above me; however, the blustery weather actually was working for me, since the rough lake prevented the floats from forming suction on the water's surface. On glassy water, I'd learned to tilt the wings slightly and lift one float off at a time, breaking the suction and allowing freedom for takeoff.

With waves of apprehension, I climbed out over the lake. Almost immediately I encountered light rain and chop. As I turned to a down-wind pattern, rain increased and visibility rapidly deteriorated. The plane seemed so small and fragile. *How had I gotten myself into this?* My first flight in a barnstorming plane at age 15 over the suntanned Kansas wheat fields had been fun — just as I thought learning to fly would be. Not death-inviting like this!

I turned to the crosswind leg and squinted through the rain-splattered windshield, searching for the green light from the tower. Only about 10 percent of the planes had radios, so most of the communication was by light gun signals, both for airplanes on the ground or water, or planes in flight. The color of light (green, red, white) and the type of signal (steady, flashing, or alternating) was a language pilots learned. The controllers in the flight tower flashed the lights precisely at a specific aircraft. For example, a red light indicated the pilot was not cleared to land and should circle until it was safe to do so, and a green light signaled the pilot to proceed to touch down. (https://www.faasafety.gov/files/gslac/courses/content/25/181/light%20gun%20signals.pdf)

At last, the tower flashed the welcoming green light and I cautiously made my approach.

Just as in takeoff, the water's surface again advised me. At least I didn't have to worry about a mirror-clear lake and lack of depth perception. The stormy waves made judging the distance easier.

I glanced over to the shoreline for further reference. Searching for the water's surface, I let the plane settle with its nose up. Even stalling a few feet off the surface would jolt the pilot since shocks didn't accompany the floats.

To my relief, the floats flawlessly skimmed over the lake. Then, remembering my fellow student's fate, I kept the stick buried in my stomach until I came off the step and the floats settled into the water. I wiped my damp forehead and taxied back to my instructor.

"Fine job. I knew you could do it!" my instructor slapped me on the back. "You may practice solo from now on."

"Not today, thanks." I said with a sigh.

I taxied back to my tie-downs. Only two weeks before, I couldn't even start this plane. Now I had just taken off and landed by myself. I couldn't believe it.

Four weeks later and with 40 more hours I went for my ticket and passed the exam with100 percent.

I had no idea that my soloing experience of soupy weather, fast prayer, trepidation, dry mouth, and damp forehead would typify many of my flying experiences to come, and that someday I would be a real bush pilot with over 3,000 hours. All I knew now was that I'd made it: Flying or Bust!

Now what was that Wally had said about hunting with the J-3?

Physicians in rural parts of Alaska

Years later, Elmer Gaede became a certified Senior Medical Flight Examiner. In the late 1980s, he figured that at least three-fourths of the physicians in the rural parts of Alaska were pilots.

CHAPTER 4

THANKSGIVING DAY MOOSE

November 1956

IN EARLY NOVEMBER, I received my land rating with my J-3, so I could not only land with floats but with wheels and skis. In summer, Alaska is one big landing strip of lakes for a floatplane, but in the snow-covered winter, skis are the ticket to landing anywhere. The J-3 now sported handsome wooden skis.

"Well, Paul, what do you say we try a moose hunt?" I asked my good friend Paul Carlson as we sat in our communal living room amidst the chatter of children playing after supper.

Paul was a good friend, and I was glad I'd found him.

Our family wasn't the only one to move into Anchorage and leave behind familiar and comfortable patterns of life in exchange for hoped-for opportunities in the Alaska Territory. Shortly after we'd moved into our rental house, a smiling, grandmotherly woman knocked on our door and extended a basket. The hostess, just like a settler in a Conestoga wagon, who welcomed travelers to a western frontier town with water and basic provisions, greeted us with a city map, newspaper, advertisements for haircuts and plumbing needs, samples of laundry soap, hard candy, hand lotion, and a collection of nails.

"Where are you from?" she asked.

When she learned we were from the plains of Kansas, she paused and nodded her head. "Hmm. A bit different here?"

Climate and geography were only the obvious. There was a lack of sidewalks for the girls to roller-skate, a clothesline for Ruby, and the absence of close ties to medical students and their wives — who had supported one another in achieving the goals of graduation, surviving with meager finances, and searching for a job once the diploma had been awarded.

Then, too, we'd left a close-knit Mennonite farming community where nearly everyone was related and where lives blended in wheat harvests, funerals, and homemade ice cream. I'd heard Ruby wonder aloud, "How do you think the Friesens are doing with his new job?" "I'd like to have someone over for Ruth's birthday, but we don't know anyone." "I sure wouldn't mind some roasting ears." I had to admit, I wouldn't have minded, either, having fresh sweet corn from her parents' field.

"A good place to make friends is in a church," our hostess said gently, giving Ruby's arm a lingering pat.

In Kansas City, we'd walked several blocks to the Covenant Church. Here, we hadn't found any church *that* near to where we lived, but we'd visited a few places and settled into Bethany Baptist Church, a primarily Swedish congregation. Paul and Irene Carlson attended there and we enjoyed spending time with them, even though our life stages differed. They were older and had a daughter, Nancy, in college, "Outside" in Minnesota. (We learned immediately that "Outside" was a term Alaskans use for anywhere outside Alaska, but usually in reference to the Lower 48 States.) Paul now worked in a Piggly Wiggly grocery store and Irene, a home economics teacher, taught classes at the Community College and at the Elmendorf Air Force Base grade school. Whereas he was easy-going and a good-natured tease, she was precise in whatever she cooked, baked, or wore.

Irene and Paul Carlson

The Carlsons had financial debt from Paul's previous work in logging and trucking, and I was paying off my medical school debt. Throughout our conversations, we recognized the opportunity to save money by living together. They owned a house at 2505 Oak Street, near Lake Otis Road, and four blocks south of the one we currently rented. (Later the address was changed to 2433.) Ruth's kindergarten would be a half-block walk to the Lutheran Church, and Naomi, whose first-grade class would be meeting in the Bethany Baptist church basement, would have a short walk of four to five blocks.

Paul and Irene turned their one-car garage into a bedroom for themselves, then big-heartedly invited our family to use the two bedrooms of the tiny house. We all shared one bathroom, the alcove of a kitchen, and a crowded living-dining room. Paul supplemented our group's food budget with broken packages of macaroni, rice, flour, sugar, and cake mixes.

While Paul and I formulated a plan for a moose hunt, Naomi and Ruth mooed and snorted with their toy farm animals beneath the dining room table. They had fond memories of Ruby's parents, Sol and Bertha Leppke, and their farm outside Peabody, Kansas. At my feet, Mark banged unremittingly on a pegboard. At this point, we didn't realize he was developing his musical abilities and perhaps even now, honing his sense of rhythm. We would embrace and appreciate this much more when he was older. Behind this din, dishes clattered as Ruby and Irene cleaned up the kitchen.

"Okay, Doc. I've been up with you a few times around Anchorage, and I think I'm brave enough to try a hunt," replied Paul, with the ever-present twinkle in his eyes. One of the things I liked about Paul was his ready-to-go attitude and love-of-life personality.

Our backgrounds didn't lend themselves to big game hunting. Even though Paul had grown up in Minnesota, neither of us had ever hunted anything larger than a coyote or jackrabbit. We each had a 30.06 and I'd just bought a 300 H & H. After several sessions of target practice, along with taking in solicited and unsolicited hunting advice, we planned our moose hunt. As was to become my pattern, my initial hunting preparation was well thought-out.

We acquainted ourselves with hunting regulations and studied the maps of various hunting areas. I penciled out a list of what to bring. On Thanksgiving Day, 1956, we took off from the snow-packed surface of Lake Hood in the crisp 0° sunrise. I pointed the plane nose south toward the Kenai (KEEN-eye) Peninsula.

I knew one rule of small plane flying was to cross a body of water at its narrowest point and to gain a safety margin of altitude in case of engine failure. I figured that at 2,000 feet, I could glide across the Cook Inlet if an emergency arose, so I started my ascent.

The J-3 took its time air-pedaling but finally got us across the gray water. The northern half of the Kenai Peninsula was a myriad of lakes surrounded by spruce trees. The ice on these lakes was probably a foot thick, covered with a layer of puffy snow. Seasoned Alaska hunters assured us the moose would be on the lakes, so we dropped to a few hundred feet to see for ourselves. Sure enough, on the second lake, we spotted three large moose — although we weren't sure in our judgment of what constituted *small* or *large*. In any case, these critters were much bigger than the dairy cows I was acquainted with.

I turned my head and called to Paul in the backseat, "This is going to be easier than I expected." I cut the throttle, and we silently drifted down to the lake.

"Yes, our moose is waiting for us," he replied with boyish enthusiasm.

The landing was feather-light, but much to our vexation, the moose were gone. They had fled into the woods. We were puzzled. Then we surmised the obvious; moose were wild game, unlike cattle, and would have to be stalked. Seeing nothing to stalk, we took off with a cloud of snow bursting behind us to resume our search for other moose.

After a half-hour, we spotted a lone bull on the edge of a small lake. With Moose Hunting Lesson #1 firmly in my mind, I decided to land on an adjacent lake and stalk the moose through the woods.

This lake did not resemble the one I'd landed on previously. Unlike the first snow-cushioned lake, which had been protected from the wind, this black-streaked lake was windblown with glare ice. All the same, as an innocent, inexperienced new ski pilot, I landed as usual. My confidence evaporated immediately. Even with the propeller completely stopped in front of us, the plane hurtled across the lake. There was no snowy resistance or drag to stop us. The trees at the edge of the lake loomed taller and taller.

We sat speechless in our seats, comprehending the danger yet not believing this was happening. My mind raced, trying to remember formal flying instructions. Desperately, I turned the plane sideways, hoping to develop more ski friction. No such luck.

"Hang on!" I yelled. Then we hit the slightly inclined shoreline. I fully expected us to nose over. Instead, we bounced to a stop. Only one wing tip nudged the brush next to the trees.

My heartbeat reverberated in my ears. We sat in stunned silence. Finally, I cleared my throat and ventured a question, "Are you still brave enough to fly with me?"

Paul responded without hesitation, "Well, we made it, didn't we?"

I turned around cautiously. He was grinning.

We unfolded ourselves from inside the plane, grabbed our guns, and crunched on the hard-packed snowy shoreline toward the other lake — where we knew we'd find our Thanksgiving moose.

In a short time, we emerged on the other side of the snow-hushed woods, once again expecting to see our moose waiting for us. No moose. No sound. No movement. Then as we were about to go back, I caught a slight movement out of the corner of my eye and saw, rather than heard, our intended prize running through the woods. It was an incredible sight to see the full-racked animal lumbering through the dense forest with nary a sound.

Sensing danger, the enormous animal stopped broadside at 200 yards and looked at us with his massive nose, gigantic ears, and beady eyes. This was definitely not a coyote or a jackrabbit. He stopped in his tracks and his hesitation gave us the edge, and at 150 yards, I aimed and fired — one shot. The monster dropped to the ground with a crash.

Adrenalin coursed through my veins, but I couldn't move.

"Do you think he's really dead?" I asked in choked disbelief.

"There's only one way to find out," Paul declared in a loud whisper.

We ran through the ankle-deep snow, around stumps and windfall, and over the uneven snowy ground toward the moose. Then, nearly knocking each other over, we stopped in our tracks. What if the beast was unexpectedly resurrected? We edged closer and warily poked it with our guns. He remained motionless.

"We did it!" We hollered in unison, slapping each other on the back, and laughing in relief and astonishment.

Elmer's first moose, Kenai Peninsula, November 1956

Then, with sobering reality, we stared at each other. *What do we do now?* was reflected on both our faces.

"It must be like butchering a steer," I suggested. I studied the large animal at my feet and then pulled out my hunting knife to cut its throat and drain the blood. He must have weighed at least a thousand pounds. I looked up around me, instinctively expecting to see a barn rafter with block and tackle, or winch, with which to hang the moose. "This isn't Kansas," I mumbled, realizing we didn't even have a rope, much less these other conveniences.

Somehow, we managed to awkwardly bleed the moose, and then farmer-like, we went about gutting and skinning the giant. I felt I was working with a plastic knife as we struggled with the tough hide.

Later I learned that, unlike farmers working with beef, which required skinning and enjoyed the sanitary conditions of hanging in a barn, hunters would quarter the meat, leaving the hide on to protect it when it was either dragged out or packed out, meat-side up.

"Look, Doc," Paul commented when our monumental task was nearly complete. "The clouds have moved in. We'd better get back to Anchorage. We don't have a tent for the night."

A few snowflakes hit my face as I mentally added "tent" beneath "rope" on my list of things to bring along next time. Maybe a sleeping bag would come in handy, too, just in case.

"By the way, how are we going to get all this meat in the plane?" questioned Paul.

I considered that dilemma. The moose seemed to be like a fishing story — growing in size.

"Maybe we should take a hind quarter with us now, cover the rest with the hide, and come back another time," I suggested. I knew bears were in hibernation, and wolves usually wouldn't eat meat handled by humans. Paul agreed to leave it in this outdoor icebox, so huffing and puffing, we wrestled the remaining moose into a natural hollow and wrapped the hide around the meat mountain.

By the time I packed out the quarter and loaded the plane, the sky was heavy. Unlike landing, takeoff on the glare ice was relatively easy.

After that, nothing was simple. Nervous tension compounded our stress. Nearing the four-mile stretch across Turnagain Arm, at the northern end of Cook Inlet, I wondered if our hunting adventure would have a "happily ever after" conclusion. The barriers between us and our destination increased. When we crossed the inlet, lowering clouds pressed us toward the beckoning murky water. A strong headwind pushed us away from the Anchorage shoreline. Darkness shrouded Anchorage, allowing only patches of landmark lights to blink through. Added to these treacherous conditions was our heavily loaded plane. My diagnosis of these facts was that the 2,000-foot altitude was no longer an adequate safety margin.

About halfway across the inlet, I remembered the gas gauge. The J-3 was an unsophisticated aircraft with a simple gas gauge — a bobbing cork. The bobbing cork within the 13-gallon nose gas tank was attached to a wire, which poked up through the gas cap, near the windshield. A short wire meant, "Find gas now!" After the wire stopped bouncing, there were usually only 20 minutes of fuel left. As the wind buffeted the plane in the inky sky, I pulled out my flashlight to inspect the bobber. It wasn't floating.

Sucking in my breath sharply, I leaned back toward Paul and confessed, "We'd better do some praying. I'm prepared to meet my Maker, but I'm not ready for a cold, wet grave."

It seemed like an eternity as we fought our way to the Anchorage shore. Once over land, we weren't home free. The Lake Hood tower was closed. We wandered through the darkness, knowing the engine could sputter to a stop at any time. At long last, I located the few channel marker lights between Lake Hood and Lake Spenard. The J-3 had no landing lights. I could only hope and pray that there were no other planes in my pathway. Slowly and painstakingly, I eased into the blackness between the dim markers. The skis reached down, and with a bounce, found the hard-packed snow. I squeezed my eyes tiredly and let out a deep breath.

Very slowly, I felt my way along as we taxied to my tie-downs. I knew we'd been flying on fumes and a prayer, but not until I shut down the throttle did the plane engine stop.

We sat in tired silence. My shoulders were tight. Before either of us moved, we both mumbled thanks to the good Lord for safety. It had been a long day. Apparently, our numbered days on earth weren't over.

I stiffly climbed out of the plane in the cold darkness. The solid ground beneath my rubbery legs felt great. Without much conversation we moved our first fruits of Thanksgiving moose out of the plane and into the car and headed home. By the time we walked through the door, the gleam in Paul's eyes had returned.

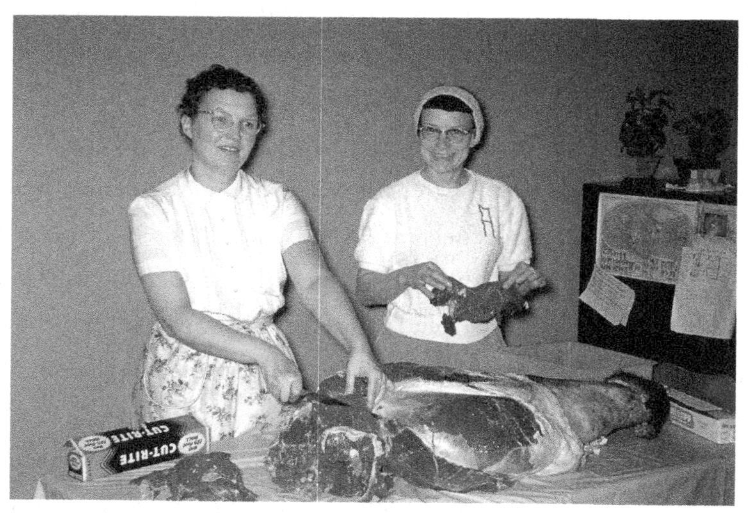

Irene Carlson and Ruby cutting up the moose, November 1956

Over the next month, we nearly wore out the first half of our hunting story, recounting the adventure to anybody who would listen. Finally, I embarked on the last half of the story. This time, I wrote down the tale and shared it with my parents in Reedley, California. Their local paper published the story:

> *Last Saturday Paul and I planned to fly out to get the rest of our moose, which was across the inlet and about 50 miles away from here. I loaded the plane with tools to get the moose out and waited for Paul to return from work, but since he didn't come by 10 a.m., I took off alone.*

About 30 miles out, ice fog began closing in from the Cook Inlet. I stayed on the edge of the fog and landed on the lake where we had left the moose. I found the moose frozen solid and in good condition. For the next several hours, I chopped and sawed apart the three remaining quarters and backpacked them to the plane. Even though my task was accomplished, I could not take off since fog had closed in around the lake within an hour after I landed.

Freezing rain then started and soon the plane was iced over. I tried once to get off the lake, but fog surrounded the entire area. Knowing I was grounded for the night, I found some dry wood, built a good fire, and tried to stay dry. I put my sleeping bag over my snowshoes and tried to sleep under the wings of the plane. It got dark at 4 p.m., so it was a long night.

The next day was just as foggy as the day before, but on my third attempt to fly out I managed to fly between fog layers over to the inlet, but I could not get across. From my low altitude, I looked for lakes with tents or cabins for shelter. After a while, I found a rundown cabin beside a large lake and landed. Blinding snow covered the plane as I cut the engine. The cabin proved to be a godsend because it was still warm — the stove had been used only a few hours previously. I spent Sunday night alone in this warm cabin.

Monday morning at 9:30 a.m., I saw an Air Force rescue plane, a two-motored Albatross, looking for me. I had filed a flight plan and was overdue since 3 p.m. Saturday. When they spotted my plane, they dropped messages to me in a bright orange bag. They had looked for me Sunday, but couldn't see through the fog. They said the Anchorage airport was completely closed in with fog. Since I was O.K., I didn't request help. Knowing they would notify Ruby that I was all right, I could rest at ease and wait for a clearing.

Monday afternoon, an Indian trapper happened to come by the cabin, and he turned out to be one of my patients from last November. He made us pancakes and fried beaver meat. We had a good time talking about hunting, his family, and his first visit to Anchorage. He had a short-wave radio so we kept in touch with Anchorage weather.

On Tuesday, the ice fog was still hanging low and the temperature remained near zero. Finally, at 1:50 p.m. the Civil Air Administration (CAA). (In 1958, this organization was restructured and became the Federal Aviation (FAA) Agency.) The weather report stated that Anchorage airport was open, so I made my sixth attempt to come in.

The fog was down to 1,300 feet over the inlet, with large ground fog patches near Anchorage, but I made it down with no problem just before dark. I had 15 minutes of fuel left when I landed.

Later I found out that a bush pilot and our church pastor had tried to come across the inlet to find me about 30 minutes before I came in, but they were forced back by fog below 500 feet, so I certainly felt God had opened the Inlet for a short time for me. I was fortunate to come through when I did because we had heavy ice fog the next three days.

I never realized I had so many friends until I came back. Church, hospital personnel, and even patients showed a great concern over my safety. While I was stuck across the inlet, some of the church women stayed with Ruby to help pass the time. After the Air Force plane made contact with me, they notified her of my safety.

Last night we thawed out a front quarter of moose and cut and wrapped it. We're having a big moose this coming Friday evening with two medical families."

And so, the second half of the Thanksgiving Day Moose story *did* have a happy ending, and the "hunters lived happily ever after." At least one hunter was wiser and bought two down-filled army sleeping bags, which were good for minus 40° to 50°— for who knew what lay ahead?

CHAPTER 5

LIME VILLAGE EVACUATION

January 1957

SNUGGLED BETWEEN the white blanket of clouds above and the bed of snow below, we flew above the Stony River searching for Lime Village. Mike, the pilot of the Cessna 180, maintained an altitude of 2,000 feet while looking out his side window into the afternoon overcast. I rubbed my army green wool mitten against the window and tried to detect any distinguishing marks against the black and white environment. Nothing but sameness.

This reminded me of my trip to Noatak with Tommy, where the whiteness spread like a never-ending roll of tissue paper in front of us. There was one difference — Tommy was an experienced bush pilot, familiar with the country and his destination, whereas neither Mike nor I had much experience in bush flying or in this hide-and-go-seek winter village-finding.

My eyes continued to scan the riverbank, while my thoughts drifted back to earlier that morning when I'd walked into the hospital, expecting a routine day. Before I could even take off my coat, the hospital medical director cornered me.

"Didn't you volunteer to fly out for medical emergencies?"

I'd quick-fired back, "Sure! Where's the emergency?"

Now, as the riverbank became a repeating pattern of white snow and black-green spruce trees, and the gas indicator subtracted the green area and moved toward the red, and the minutes added up into another half hour, I asked the same question: Where's the emergency? Hoping it wouldn't be us in a downed plane.

The urgent message had come that morning via radio through the Sparrevohn Air Force Base, which was located across the Cook Inlet and through the Alaska Mountain Range:

"A baby very ill in Lime Village. Will probably need to be evacuated. Villagers will mark area on river near village where you can land. Come quickly."

At this moment, Mike and I were searching for that small Dena'ina (Duh-NY-nuh) Athabascan village with improvised airstrip.

I'd found Mike at Merrill Field where I checked for a charter plane. Due to liability issues, even when physicians, such as myself, were pilots and had their own airplanes, Public Health paid charter pilots for medical flights. The thin young man with pale eyes listened non-commenting when I explained my situation and the immediacy of departure. He wasn't one for long conversations and in about two sentences I learned he'd been in Alaska slightly longer than I, and had taken flying lessons upon arrival. He didn't exude much confidence and I deduced quickly that he was a low-time pilot who was trying to build up hours. All the same, my choices had narrowed. He *was* willing *and* he had a Cessna with a wheel-ski combination. Ironically, his plane was actually better equipped for the bush landing than he was.

My eagerness to get going blurred my good judgment. I was convinced that even with our mutual lack of experience, our determination would get us to our destination. It wasn't until we were 200 miles west of Anchorage, following Stony River and without a hint of civilization that I realized we were truly the blind leading the blind.

Together we'd studied the aeronautical charts to locate Lime Village. One chart we looked at had a village called "Hungry" marked where Lime Village should have been. Later I learned that someone had jokingly called the village "Hungry" and the name stuck.

We traced our route from Anchorage, west through Merrill Pass in the Alaskan Mountain Range, to Stony River. We would follow Stony River to Lime Village, in the Lime Hills, and on the south side of the river.

"I'm sure we should have been there by now," I said with some exasperation. "Let's turn around, drop down to 500 feet, and backtrack."

About ten minutes later, I recognized smoke spiraling upward from what appeared to be a mound of rocks. We circled low and people immediately popped out from the now-perceptible dozen cabins. Another several rounds and children waved, some men pointed upriver toward the east, and other men hitched up their dog teams.

We banked the plane, and about a mile upriver we discovered spruce branches marking our runway. Mike made several exploratory passes. "What do you think, Dr. Gaede?" he asked, keeping his eyes on the landscape below.

"It looks okay to me," I stated, in spite of my apprehension. I realized he'd never landed on anything like this before. "Let's give it a try."

I tugged at my seatbelt, and looked out the window to be sure he'd pulled the skis into position and the wheels out.

Mike pulled full flaps as we flared down for a ski landing. So far, so good. I thought we'd made it, but in a flash realized we hadn't. The rough ice hit the skis violently, throwing us around in our seats. I felt as though I was at the end of a jackhammer.

The plane skidded and clattered around on the airstrip. I gripped my seat edges. Eventually we slammed against a short pressure ridge and stopped. We sat there for a long second. Mike finally had the where-with-all to shut down the engine. The plane shuddered into silence. In unison we let out huge gasps. We mirrored each other's amazement.

"Do you still have your teeth?" Mike asked.

The guy did have a sense of humor.

He looked out the plane window and declared with more animation that I'd witnessed the short time I'd known him, "I'll *never* let anyone choose my landing spot again!"

I shook my head, "Things certainly aren't the same on the ground as they look from the air."

We got out of the plane and examined the skis.

"We're lucky these skis didn't tear apart!" Mike declared.

Just as we finished pulling the cowling cover over the engine to keep it warm, a group of dogsleds arrived. I'd seen the dog races at the Fur Rendezvous, but this was my first time to see *real* dog teams up close. The dogs looked quite bony beneath their fur. Their masters, who greeted us with friendly smiles, were none too heavy either. I wondered if this had anything to do with why the village was nicknamed "Hungry."

The driver of one team motioned for me to get in, and another waved to Mike.

This will be great, I thought as I climbed in. I imagined gliding along, in the same way I'd seen the dogsleds do in front of our house. Already I visualized telling my family and friends that I'd actually ridden in a dogsled!

The dog team driver yelled at the dogs to "Mush" and gave the sled a push. He hung onto the sled back and ran behind until the sled picked up speed; then he jumped onto the back runners. My enthusiasm diminished as we bounced and careened between pressure ridges in the rough ice. It was a replay of our landing, except for one thing — the plane seats were cushioned. The sled bottom had no padding or shock absorption. After an eternity of 15 minutes, my misery ended and I gingerly extracted my abused body from the sled. So much for the glamour of dogsled riding.

"The baby is there," said Jacob, our designated host, pointing toward an older cabin a short distance away. Mike and I followed him slowly along a narrow snow path. Some of the children trailed behind us, trying to hide behind one another and smiling shyly. When I smiled back, several giggled, ran up and touched my arm bravely, then ran back into the group.

A typical log cabin at Lime Village, November 1956

Once at the cabin, we kicked the snow off our boots. The doorway was framed by icicles stretching from the roof nearly to the ground. A stuffy warm heat enveloped us when we entered the dimly lit 20- by 24-foot log cabin. Mike remained standing to one side, looking like he wasn't sure what to do with himself.

The anemic January sun filtering through the dull sky passively stayed outside the small smudged windows. Inside, the only light flickered from a kerosene lamp sitting on a wooden table, amidst a jumble of powdered milk, dishes, a pocket knife, wood shavings, and a raw red fox hide. A fat 55-gallon oil barrel stove produced the heavy heat in the room. The fuel drum was turned on its side and set on metal legs. A hole was cut for a door at one end and a stove pipe inserted at the other. An attached piece of metal, welded to the top, provided a flat surface on which to heat water and food.

On a bed in one corner sat a young Native woman nursing her baby.

"Jean, this is Dr. Gaede," said Jacob.

"Doctor, my baby John has cold a long time," said Jean.

I pulled up a chair near her.

"Now he coughing more and more. He has high fever. He don't eat good. He getting skinny." Her concern was evident. "He don't act right. Others in village cough much, too, but they not sick like my baby John."

I examined the frail baby who had a feeble cry and a weak productive cough. He was pale, in spite of his Indian-brown skin, and I guessed he was anemic. He had a high fever, some evidence of dehydration, and moisture in both lungs. I mentally added up these symptoms.

"Jean, your baby is very sick and probably has pneumonia," I gently explained. "I'll have to take him with me to the hospital in Anchorage — or else..."

"Doctor, you take him now," Jean quickly replied as though she had thought through the alternatives. Within minutes, she had gathered some baby clothes, wrapped more blankets around him, and handed him to me. I cuddled the tiny, unresponsive child. Jean walked us to the door, but before opening it, stopped me, pulled the worn flannel blanket away from baby John's face, and kissed him on the forehead. I felt bad for her, but there weren't other options.

I was not eager to take another dogsled ride, although with the quickly fading winter afternoon daylight, I knew my only other alternative — walking back to the plane — would detain us from the time-demanded takeoff.

When we got back to the strip, I passed the baby to Jacob while Mike and I walked around the runway to determine the least perilous path for takeoff. The conversation was short and we did a better job on our takeoff course. Soon we were heading back to Anchorage. Just as the sun fell out of sight, leaving a pink, deepening to purple afterglow against the Chugach Mountains, we touched down.

Mike and I congratulated ourselves on our bush-flying success. I thanked him and wished him well, then headed to the hospital with baby John.

X-rays and a laboratory workup showed that the baby not only had pneumonia, but also active tuberculosis. He would have to stay in the hospital many months before returning home.

Two months later, the Anchorage ANS hospital received a similar emergency radio call from Sparrevohn Air Force Base. Apparently, a dog team from Lime Village had arrived with another desperate plea for medical help. This time it was for a two-year-old child, who was very ill with possible pneumonia. Could medical help be sent as soon as possible?

"Elmer, are you game for another mercy flight?" my supervisor asked me.

"Sure," I replied. "I'll inquire about a charter plane."

A pilot and plane were available the next day and even the weather was forecast to cooperate. Rather than flying into Lime Village, I would fly into Sparrevohn, 20 air miles south of Lime Village. The sick child would be brought in on dogsled. I thought about my own recent dog sled experience and felt sorry for the child already. I couldn't imagine 20 miles on a dogsled. Maybe I was just too tender and the Natives were used to the bumping and jostling.

This time, my pilot, Dennis, was an experienced bush pilot and already acquainted with flying through the mountains and landing at Sparrevohn. He was older, and if gray hair indicated anything, he was wiser — or at least more experienced than my previous charter pilot. I didn't have to nudge this pilot along.

Sparrevohn was a restricted Air Force airfield, and authorization was required. Dennis took charge and called ahead to request permission to land on the 3,700-foot runway. The Civil Aeronautics Authority (CAA) weather report indicated early fog in the mountains and in low areas, which was to dissipate as the day progressed. As critical as the young patient was, we didn't fire up the plane until noon. Sparrevohn was giving out reports of sky obscured and visibility of less than a quarter-mile. This was expected to change by the time we got there.

The flight route again took us west, over the Cook Inlet. As we approached the Alaskan Mountain Range, wide bands of blue-gray clouds allowed only a partial view of Merrill Pass. These we skirted and then flew above the fog patches crouched in the Pass valley to Stony River. We then cut over to Cairn Mountain.

Although we had flown over only minute patches of fog along our route, we were disappointed to find the entire valley at Sparrevohn socked in, with no visibility of the ground from the air.

The Sparrevohn Air Force Base fixed radio operator informed us that the fog was rolling in and out so that occasionally the runway was visible.

"Go into a holding pattern to wait for a break in the fog," he advised.

"We can try making a special IFR (Instrument Flight Rules) approach," Dennis calculated out loud.

We'd discussed the airstrip earlier. I knew there would be only one attempt — with no mistakes. The runway ended at the base of the 3,800-foot

Cairn Mountain on the north. Since the runway was elevated on the north end, all landings were made to the north and all takeoffs to the south.

I had never landed in below-minimal VFR (Visual Flight Rules) conditions and felt apprehensive. At least I had Dennis and not Mike. I wouldn't have been able to lend any authority on the circumstances.

We circled about 30 minutes, after which Dennis turned to me and said, "My fuel is nearly half gone. If we can't find an opening in the fog soon, we're going back to Anchorage. I'll talk to the radio operator once more and ask about the thin spots."

Through the cloud base of 500 to 800 feet, we could intermittently see patches of ground.

Just as he was reaching for his microphone to call the operator, he blurted, "Hey, there's a hole! South end of the runway — perfect. Let's give it a try."

We slipped the plane, making a quick descent using opposite aileron (wing) and rudder (tail) control through the hole and pulled flaps, squeezing under the fog, where we could see nearly all the runway. I was still trying to get used to the dropping feeling this caused, especially as a passenger, rather than a pilot. Added to the stomach-turning thrill was our low altitude. We were below the normal approach pattern and just skimmed the terrain, before dropping down on the numbers of the runway.

We touched down and rolled past a wrecked military cargo plane to my right. Yes, this was another hazardous landing. Were there such things as *normal* bush landings, I wondered?

An Air Force officer met us and took us to a nearby barrack. We found not only one sick child from Lime Village, but two — both boys between the ages of two and three. They cooperated with my examination. Neither one smiled or giggled, as had the other children I'd met at Lime Village. Instead, they sat listlessly, flushed with fevers, and coughing. I was betting on pneumonia.

"Men, it looks as though you're right," I said, turning to the two dogsled drivers. "These children are very sick and I'll need to take them both back to Anchorage."

The men nodded stoically and handed me bags with the children's clothing.

"We afraid our babies would die. Go now."

They solemnly patted the children's heads and walked out into the cold. The children didn't say a word, but held each other's hands. Dennis followed the men and looked up at the sky.

"There are a few good holes up there. Let's get out now before any more trouble rolls in."

Complacently, the boys crawled into the back seat of the plane. Unlike healthy children, who would be absorbed by the novel experience, they curled against each other and fell asleep even before we taxied up to the north end of the runway.

Taking off downhill was simple, and we easily climbed through the patchy fog. We leveled out at 6,000 feet and followed our same route back, pushing through the gray passes. As we emerged from our foggy tunnel, the bright sunlight greeted us, waking the children. They pressed their faces against the windows and studied the broad expanse of murky water, followed by the silver mud flats, and then the buildings of Anchorage.

Complete physical examinations with X-rays and laboratory tests revealed active tuberculosis and pneumonia in both children. These children would eventually recover, as would baby John, but the discovery of three children in the same village with active tuberculosis caused our staff great concern.

The following summer, a tuberculosis survey team was flown to Lime Village. The nurses and laboratory technicians carried with them a portable X-ray machine. The findings were shocking. Twenty-eight of the 29

Natives had active tuberculosis. Lime Village was literally wiped out as all those with active tuberculosis had to leave the village for prolonged treatment.

Needless to say, our 400-bed hospital soon reached capacity, with not only these villagers, but others, and we had to send many Native tuberculosis patients Outside to the United States for treatment.

Later, when the Lime Village Natives could return home, many chose to make a new start and relocate to Stony River village along the Kuskokwim River.

Baby John had been just the tip of the iceberg in the unfolding saga of one village's encounter with the ravages of tuberculosis.

Tuberculosis in Alaska

Tuberculosis (TBC), which in the past was called consumption, was a historic backdrop during the years Dr. Gaede practiced medicine with the Public Health Service in Alaska. It is caused by a bacterium that leads to different levels of infection. It is believed to have been brought to Alaska by Russian and American whalers in the 18th and 19th centuries and became the most urgent and important health issue in Alaska.

Active symptoms include fever, chills, night sweats, weight and appetite loss, fatigue, chest pain, and persistent coughing. The disease can reside dormant in a person until the immune system is weakened through alcohol, malnutrition, AIDS, unsanitary conditions, or other matters. It can affect the lungs, kidneys, bones, and joints. It is transmitted through coughing, sneezing, or talking.

A medical survey in the early 1930s estimated that over one-third of Native deaths in Alaska were attributable to tuberculosis. In1946, a school teacher in Barrow reported that of 30 children who entered school between the ages of five and six, only six lived to finish. In 1949, 23 percent more people in Alaska died of tuberculosis than in the United States. When the Public Health Service took over the Bureau of Indian Affairs medical work in 1955, they found an appalling amount of TBC among the Natives. Instead of the typically less than 80 per 100,000 number of cases in the general U.S. population, the Alaskan Native's cases were 2,300 per 100,000 — nearly 30 times more prevalent.

What was found in Lime Village was also found in other villages; consequently, TBC case-finding became one of PHS's goals in 1957.

Typical treatment is a weekly regime of oral antibiotics for six months. Although this appears simple, a number of issues complicate the process. First of all, people must be diagnosed, which requires funding, personnel, and equipment. Then, they must take the medication regardless of whether they feel better, which oftentimes requires the oversight and persistence of a healthcare worker.

Dr. Gaede was in close contact with active tuberculosis for four years, but never developed a positive tuberculin test. This is nearly unheard of for that kind of exposure. Throughout his years of medical practice, his wife, Ruby, expressed frequent concern that he might carry home this and other illnesses. However, never did anyone in the family ever contract any sickness directly attributable to the patients he treated. Dr. Robert Fortuine documented this killer extensively in his articles and books, including "Must we All Die? Alaska's Enduring Struggle with Tuberculosis."

CHAPTER 6

OUT TO GET A BEAR RUG

March 1957

WE'D HEARD OF HIM, the Brown Bear who stalked hunters. Just the previous year, a lone hunter tracked him, but never returned with a story, much less a rug, and no one ever found the man. In the emergency room, I'd patched together the mangled bodies of men who had encountered similar killers.

Brown bears, or grizzlies, are a favorite topic with most hunters, hikers, and fishermen. The brown bear closely resembles its relative, the black bear, except that it is larger and has a more prominent shoulder hump — as a result of muscle built up from digging. Its claws are not retractable, making its paw prints distinct and its viciousness more severe. Color in itself is not reliable since both the brown and black bear come in varying shades of black-brown.

Some hunters seek bear trophies in the fall, before the bears enter dormancy in November. Jim Orr and I decided to wait until spring. Jim, an employee at Elmendorf Air Force Base, had lived next door to our first rented house. He, his wife, and teenagers were a "welcome wagon" of their own with their neighborliness and tips for finding our way around the community. On many occasions, he told hunting stories, which kindled my interest and carried me into my first hunting experiences. In his early forties, he looked every bit the man's man, well-built, fit, self-confident, outdoor-savvy, and capable of handling himself in the face of any adversity. He was a seasoned Alaskan; whereas I was a Cheechako (Chee-CHA-koh), as Alaskans referred to us naïve newcomers, and which was at the other end of the spectrum of a sourdough. And, I was a novice pilot. In other words, a prime candidate for trouble and in need of a mentor. My ambitions, nonetheless, equaled those of gold-seeking optimists compelled to take advantage of this rich territory and make their wildest dreams come true.

March 30, 1956 marked my introduction to bear hunting. At 6 a.m., in 30° temperatures, Jim and I loaded my ice-glazed J-3 on skis at Lake Hood. I was an amateur and minimized the lift-reducing effects that frost, much less ice, had on a plane. Fortunately, however, even with Jim's stocky 200 pounds, the plane crawled off the ice-covered lake. The higher altitude, rising sun, and wind melted the plane ice as we headed north over a narrow finger of Cook Inlet and toward the Talkeetna (Tal-KEET-na) Mountains.

Before us, Mount McKinley and Mount Foraker stood as glittering landmarks in the clear horizon, their peaks bathed with thin sunlit clouds. I'd never met mountains that were a *couple*. Mount McKinley is referred to by the male pronoun with the connecting mountain, Mount Foraker distinguished as "his wife," and subsequently regarded by the female pronoun.

I'd carefully planned this trip step by step and filed a flight plan, which took us first to Willow, 40 miles from Anchorage. There we followed the

Alaska Railroad for another 40 miles to Talkeetna, at the base of the Alaska and Talkeetna mountain ranges. Appropriately named by the Indians, "where the rivers meet," the small one-street town with miscellaneous log cabins and a roadhouse or two was located at the confluence of the Susitna (Soo-SIT-nuh), Talkeetna, and Chulitna (Chew-LIT-nah) rivers.

We flew east, into the Talkeetna Mountains, to a ridge where Jim had successfully spotted brown bear in previous years. Skirting the ridges at 3,000 feet, we saw moose, but no signs of bear. Only the dark shadow of the plane flitted against the still background.

"Doc, I think we're a few weeks early," declared Jim, leaning forward and speaking loudly over my shoulder.

At this same moment, I looked out the front window at my gas gauge wire. I decided to stop at Talkeetna to refuel before exploring further.

While fueling, I suggested to Jim, "Let's try it to the west." I wasn't ready to admit defeat so quickly on my first bear hunt.

We flew west in the clear, dark blue sky toward Mount Yenlo in the Alaskan Range. Suddenly, after canvassing the valleys and ridges, I caught sight of distinct, fresh bear tracks, black against the white ridge. Unlike moose tracks, which followed a gradual ascent along the mountainside, these tracks shot in a straight line, up one ridge and down the other. As we followed the tracks, we noticed that the bear actually slid down the ridges as though they were great playground slides.

After a short distance, the tracks strayed away from the mountains to the open country, where only an occasional bush pierced the unmarred snow. I looked ahead, certain we would easily find him in this unprotected area.

Then I spotted him, at first only a brownish-black object at the end of the trail. Spiraling closer, we dropped to less than 100 feet above the moving mass, and made several passes. He tried to hide, then finding no seclusion, jumped toward us in challenge. Standing nearly nine feet tall on his hind feet, he clawed at the air beneath us, his mouth open in rage. Here he was, the Alaska brown bear and the largest meat-eating animal that lives on land. He reminded me of a 1,200- to 1,400-pound Kansas bull.

In the northland, every sportsman yearns to have a bear rug trophy. "This bear obviously does not aspire for the glories of the hearth-side," I shouted to Jim above the engine's roar.

"No, but I think he's in for a surprise," Jim confidently yelled back. "Let's put this thing down and get what we came for."

Over a small hill, I spotted a flat, smooth, potential landing area about two blocks square and only three blocks from the bear. We made several more passes over the bear, attempting to nag him closer to our landing spot.

"Keep your eye on him!" I said. "I'm going to land."

As soon as we touched the unbroken white carpet, I felt a tremendous drag. I was in trouble. In my assessment for a landing strip, I'd checked for smoothness, not snow condition. I immediately applied full throttle, attempting to free myself from the clinging snow, and take off again. Regardless of my efforts, we gradually lost speed and even with the engine protesting, the plane wallowed to a stop. My hopes sank, as did the tail, which settled into the deep snow.

Later I learned that before committing oneself to landing, it was wise to first do a touch-and-go, where the plane skis would just skim the snow's surface to test the softness and depth, without landing.

Ignorantly, I assured Jim that we could worry about takeoff later, after we got the bear. I foolishly failed to consider the chances of the killer brownie turning on us. In this open valley with only sporadic clumps of trees, where could we find safety? A retreat to our plane with its thin fabric would only tease the short-tempered bear, which, with a single swat of a front paw, could kill a caribou or moose.

In spite of these facts, we fastened on our bear-paw snowshoes and checked our weapons. We were positive that between us and danger stood Jim's .300 Magnum, and my 300 H&H.

Walking on snowshoes demanded concentration. Practicing in the gravel pit across from our house, I'd quickly learned the consequences of not keeping the toes up. Before I knew it, I'd toed down, caught a tip in the snow and plunged headfirst into the snow. Or I'd walk with my legs too close together, the insides of the snowshoes stepping on top of one another, then dipping toward the center and causing my knees to knock together.

I followed Jim, my experienced guide. Somewhere before us was a killer bear, and behind us trailed a messy snowshoe path to my plane which was stuck in the snow. The scene was something that could have come out of a Laurel and Hardy movie. In this precarious style, we approached the area where we had last seen the bear.

Fortunately, we were downwind. Jim hunched down and then poked his head over the crest. Standing up for a moment and looking all around, he called over his shoulder, "He's gone, but boy what a trail he left for us."

Twelve-inch paw tracks four feet apart in the deep snow showed us the bear's haste in getting away and left us a definitely marked trail. Through our rifle scopes, we followed his trail for about a half mile toward the mountains, hoping to find him in our sights. No such luck.

Trudging after our trophy, we soon removed our coats as we sweated from the exertion and the sunshine reflecting off the snow. Within a half hour our efforts were rewarded, and we noticed a dark spot against the glaring whiteness. Even with the rifle scope, this appeared to be only a black ball. Drawing closer, we finally saw the outline of a brownie. Jim advised me to stay by a grove of trees and that we should try our shots with a solid gun support.

The bear, uphill and sitting down, appeared to be about 400 yards away. We balanced our rifles in limbs of trees, placed the scope cross hair right below the bear's shoulder, and began shooting. One. Two. Three...Six. The bear casually turned to look at us. After a few more shots, he growled in annoyance, turned, and resumed his journey up the mountain. As he reached the ridge, we threw more lead at him.

We hustled up the hill after him and examined the spot where we had disturbed his sunning. There we saw our shots had not even tickled his toes and had only disturbed the snow a few feet below him. We'd misjudged the distance by 200 yards and the drop of our bullets was more than we'd expected.

By now, the arduous trek had caught up with our enthusiasm. "Let's go see what we can do about the plane," I sighed.

Our exercise continued as we sweated and tramped out a 600-foot long by 12-foot-wide runway with our snowshoes. Once we'd dug out the plane and placed the tail on solidly packed snow, it was easy to push around, and we used it to further compact the makeshift runway.

Only one problem remained: a ravine abruptly ended the runway. "Jim, you'll have to tell me immediately when we are airborne, otherwise I'll need to chop the throttle at the end of the runway." We had our work cut out for us.

Giving the plane full power, I began gliding across the packed snow. I coaxed the tail off the runway. The airspeed indicator hesitantly edged upwards as we neared the end of the runway. Slowly I pulled the stick back at 45 mph and we said goodbye to the snowfield.

"Doc, you only had inches to spare," gasped Jim.

But we'd made it and I was focused on the bear. We easily tracked it by air. His trail ambled across two valleys and returned to the ridge where we had seen his original tracks. Apparently, his den was nearby.

Tired, sunburned from the reflection off the snow, yet still excited, we landed on the late afternoon slush of Lake Hood. We were already planning our next hunt.

Jim and I returned to the Talkeetna Mountains a second and third time. No success. We were determined to bring home Brown Beauty and made a fourth attempt. With the middle of April "breakup," Lake Hood would soon become too dangerous for takeoff. This was my last chance.

Alaskans do not say "springtime," instead they refer to "breakup." This signifies the time of year when river and lake ice breaks up and when the snow melts, yet the ground below is still frozen. The unsightly and messy result is abruptly formed dirty lakes in parking areas, streets, backyards, and landing strips. This is accompanied by daytime melting and nighttime freezing. Yet, it is a time for hope of warmer days and the return of a colored palette following months of whiteness.

Just that morning, we'd discovered my plane in a foot of water when we arrived at 5 a.m. Unperturbed, we had pulled the plane onto solid ice and had taken off. Like a homing pigeon, the plane easily flew back to the Talkeetna Mountains, where the previous week we'd spotted four bear dens in the area.

"It's about time these bear woke up," I hollered. I dropped altitude and flew low over the ridges, hoping the roar of the engine might serve as an alarm clock. At the same time, I wanted to keep my distance. The previous year, two hunters were killed trying to lure a grizzly from its home. It was better to allow a bear a voluntary debut.

"Look over there! There are tracks by that den," yelled Jim.

The tracks only circled the immediate area.

"Maybe the morning air is too chilly for the Sleeping Beauties to take a morning walk," I replied.

I mentally marked these dens, then flew farther into the Talkeetna Range, exploring the hills and valleys and finding six dens with evidence of activity.

"That gas bobber is kind of short," said Jim after a while. "Let's go back to Talkeetna, fuel up, and grab some lunch."

I glanced at the fuel gauge. "We're okay. Let's check once more. Maybe our trophy rugs are at the first dens."

We flew back, high over the ridges. Suddenly I shouted, "Look ahead! See the two brown bears?" The golden bears romped in their front yard, about one-third of the way up a 2,400-foot mountain. They batted at one another, rolling and tumbling in play. Chasing one another, they slid on their backs part-way down the mountainside and then collided against each other before racing back up to the top. Their play belied their dangerous natures.

"Elmer, put her down — not where we have to dig out!" urged Jim.

The ridge had several large, smooth snowfields. I put the plane down on a white bed about half a mile from our target. A firm landing. No problems this time.

Once again, we strapped on our snowshoes and made waffled tracks to a ridge overhanging our quarry. The wind was in our favor. We stopped and listened. The air was deathly silent. Did something know something we didn't? Were there eyes watching us? We peered cautiously over the loose, snowy ledge.

The bear playground, 200 yards below us, was empty. "Maybe the sunbathers had their quota of sunshine and returned home for another nap," I whispered to Jim.

"I just hope they're not somewhere watching us and considering making us playmates," he replied, clutching his rifle then standing up and looking all around.

We sat down to await new developments. After 45 minutes, Jim could take it no longer. He restlessly walked over toward a short tree about 40 feet away. About 20 feet before the tree, he dropped his rifle and pulled out his hunting knife. Optimistically, he trimmed out a gun rest.

Meanwhile, I scanned the scene around and below me with binoculars, hoping I hadn't overlooked some important factor in our hunting setup — such as a bear hunting us.

Suddenly, the two overgrown teddy bears burst into view from behind a large spruce tree near their den.

"Jim, they're back!" I yelled.

I slid next to a tree for my gun rest. Jim dashed for his gun. The bears, hearing us, started their wild scramble for safety. Jim's trigger finger quickly found its position and placed a bullet in the larger bear's shoulder, causing it to turn its head and bite the wound. His next shoulder shot threw it off balance, and it rolled into a canyon 200 feet below.

For a moment, my first shot slowed down the smaller bear, but it continued gaining speed as it started downhill. The second shot went wild. The third shot hit pay dirt, and the bear dropped in its tracks. We stood looking at each other wildly, panting from the intense rush of adrenalin.

"We did it! We got our bear rug trophies!"

In near disbelief, we hurried as fast as our snowshoes could go down the hill to our rewards.

Skinning out our prizes was tough work. Even though we'd failed to check their pelts for rubbed areas, their blond-tipped fur was even throughout. It took us nearly three hours to remove their hides, by which time we were fairly saturated with bear oil. From nose to tail, the larger bear measured eight feet and the smaller one six-and-a-half-feet. Unlike moose or caribou, brown bear meat is greasy and not something our wives would want in the kitchen. I'd heard that black bear was like pork and was palatable. I'd have to test that assumption at a later date.

Elmer's grizzly in the Talkeetna mountains, April 1957

At this moment, we hadn't eaten for ten hours and the over-exertion and adrenalin come-down hit us like a brick wall. I should have thrown in a ring of bologna, my favorite sandwich-makings; but packing along food wasn't a priority on this trip — nor was it ever. In the future, I'd keep chocolate bars in the plane.

Packing our treasures up the mountain to the plane took us step-by-step grueling hours. By this time, the mid-afternoon sun had softened the snow's crust — a threat for taking off. Packing in two full bear hides with heads, plus 120 pounds of hunting gear and my robust partner grossed our weight to the limit — and increased the sag of the tail.

I walked the ridge and measured 300 feet of good take off space before there was a steep drop-off. "We may have to drop off over the edge if we can't get enough airspeed for takeoff," I informed Jim.

"Sounds like a thriller," he replied wearily.

With full power, the J-3 slowly picked up speed on the soggy snow. The 300 feet rapidly disappeared beneath us and we dropped off the precipice, hovering in the emptiness, then, like a roller coaster, swooping down 1,000 feet before gradually leveling off. After a while, our pulses leveled off, too.

Back at Talkeetna, Don Sheldon approached us when we landed at the Don Sheldon airstrip. He was my age and already an Alaskan legend. He'd perfected high-risk glacier landings and flew teams of scientists and climbers onto Mount McKinley. Someday his feats would come out in *Wager with the Wind: The Don Sheldon Story.*

From 50 feet away, the six-foot rugged-looking man grinned, sniffed the air, and commented amiably, "Smells like you boys got your bear."

"Yep. This time we didn't just exercise the wildlife," I said. "We got our bear rug trophies."

I flew back to Anchorage in good spirits. When I'd bumped into Anchorage, I didn't know what the future would hold. Now, less than a year later, I'd learned to fly an airplane, bagged a moose, shot a bear, escorted a Native baby to the Interior of Alaska, survived a record-breaking winter, and logged all kinds of medical experiences. Rather than these adventures satiating my desires, they only served to whet my appetite for more. What could I do next?

CHAPTER 7

ASSIGNMENT: TANANA

July 1957

I SMELLED IT BEFORE I SAW IT. Smoke.

I was flying cross-country from Anchorage to Tanana (Ta-NUH-naw), a village stretched along the riverbank where the silty Tanana River lost itself in the mighty Yukon, 300 miles north of Anchorage and 130 miles west of Fairbanks. This Athabascan Indian village was accessible only by plane — or boat in the summer and dogsled in the winter.

What had prompted me to head into the Interior with my J-3, a lightweight but sturdy aircraft that was minimally equipped to fly on short sightseeing jaunts or hunting trips? It had no battery or radio. I had to carry extra cans of fuel and a fuel pump in my back seat to travel the distance. Besides these odds against me, I'd never piloted my own plane cross-country flight before.

My incentive was a new assignment at Tanana Alaska Native Service Hospital. This transfer came after completing my two-year term at the Anchorage hospital. It also followed soul-searching by Ruby and me. Since mission work was our original goal, we again felt the tug to seek out such possibilities. The Mission Covenant group presented the need for service at Nome. This appealed to both of us.

After much consideration, however, Ruby and I acknowledged that because of the nature of my work, I could follow my faith in whatever job I chose, regardless of the agency. My experience at the Anchorage Native Medical Center had been positive, and there were openings in other Alaskan Public Health hospitals. The added attraction was their remoteness — which, for me, spelled adventure.

Ruby had adjusted to Alaska, at least to Anchorage. She'd made friends, figured out how to cook with what was available in the grocery stores, and accepted moose meat as a fine alternative to beef. She expressed no objections to staying within the Territory and transferring to a more isolated setting.

I made my requests to Public Health.

First choice: Tanana, along the Yukon River, Interior Alaska, sole physician, and administrator.

Second choice: Dillingham, west in Bristol Bay above the Alaska Peninsula, two or three physicians.

Third choice: Bethel, also west, along the Kuskokwim River, four physicians.

Now the wait was over and Tanana was mine — if I could get there. I'd carefully studied the aeronautical charts and plotted my course. The village lay two miles west of the junction of the Tanana and Yukon rivers, and 130 miles west of Fairbanks. I hadn't filed an official flight plan. The radio operator at the Anchorage hospital had called the Tanana hospital notifying them to expect me this coming weekend and I'd left a map, marking out my intended route, with Ruby and Paul Carlson.

Relocating the J-3 to Tanana was only one part of the moving process. Ruby had calculated a year's worth of groceries. Then we bought $1,000 worth of staple goods from flour, sugar, coffee, shortening, and laundry detergent to canned fruits and vegetables, cereal, powdered milk, and instant potatoes. Ruby marked and addressed the stacks of boxes, and we hauled them to the railroad station. Their next stop would be Nenana (Nee-NA-nuh). From there our staples would travel by stern-wheeler barge down the Tanana River to Tanana. I was just beginning to understand the significance of the river highways that carried life to the villages that gathered along their banks.

While our supplies were finding their way to our new home, and I was relocating the J-3 and acquainting myself with the Tanana Native hospital, Ruby and the three children flew Outside to see her parents in Kansas and my parents in California. Before the fall school session, they'd arrive in Tanana.

The smoke wasn't my first roadblock — or air-block — on this venture. The day before, on a bright blue July afternoon, I'd skimmed off Lake Hood and headed north, only to find puffy white thunderclouds brewing among the mountain peaks and spilling down into the valleys, daring me to find my way through to Summit and toward Mount McKinley Park. (In 1980, the name was changed to Denali National Park and Preserve.)

I'd easily found my checkpoints, beginning with the Susitna (Sue-SIT-nah) River below me and the railroad securely to my right. Soon Talkeetna slipped beneath me. Next, Curry. The railroad cut through mountains about 4,500 feet tall, and curved around small blue mountain lakes cradled in the deep green valleys. When I came out the other side of the mountains into a flat valley, I saw boiling white clouds. I was a low-hour pilot, but I knew better than to mess with a thunderstorm. I would need Plan B. But what *was* Plan B?

I pulled out my map and considered going around the thunderstorm. Not possible. Summit with Windy Pass was the only pass to the north. I didn't want to backtrack. I looked around. With a floatplane, the lakes provided unlimited landing opportunities. There waiting for me was the answer — a 3,000-foot lake.

I dropped down and with a swoosh of sparkling spray, landed. I jumped out and into the water with my hip boots — expected attire for a float plane pilot — and towed the plane to marshy grass near an incoming stream. Pulling out my fishing rod, I cast a shiny artificial lure into the area where the icy clear water melted into the dark lake.

On my second cast, my rod jerked and the fight was on. After several minutes of playing with the fish, bringing it in, letting the rod spin out, and then bringing it in again, I pulled in what appeared to be a large rainbow trout. I measured it past my knee — about 22-inches — the largest rainbow I'd ever seen! I cast again and again. Nearly every other cast brought in a trout 20-to-24 inches long. I'd fished around Anchorage but had never encountered success like this. This was paradise! Unfortunately, I had no means to preserve my catch, and I knew I wouldn't be going anywhere soon. I released all but one fish.

Toward the north, in my air pathway, I could still see the magnificent thunderstorm, with clouds shoving against one another, pushing higher and higher in the sky. Around the dark edges of this energy pile, lightening flashed, warning me not to enter the canyon.

I'd learned from earlier trips, such as the moose hunt, to take along emergency gear; I dug around in the plane until I found a lightweight skillet. Being a Good Boy Scout, I always carried matches, and soon a merry fire crackled on the sandy shore in the early evening summer light. I chose "Rainbow Trout" on the menu and as it turned out, it was large enough that I didn't have to worry about side dishes or a dessert.

I had remembered to add "tent" to my emergency gear list, and after pushing several rocks out of the way, I pounded down the stakes. Before settling in for an overnight camp-out, I disposed of the fish bones — a good distance from my campsite. If a bear needed an appetizer or dessert, I didn't want to be on *its* menu.

Before heading into the confines of my tent, I sat beside my flickering fire, taking in the remote beauty of the evening shadows reflected on the lake framed by tall spruce trees. I felt content. My stomach was full, I had a fish tale, my floatplane gently rocked nearby, and here I was in this awesome land that exceeded my dreams. Times like this reminded me to stop and thank the Creator for His magnificent handiwork. It also made me forget the winter when the defeated sun barely hung onto the horizon and every living thing crouched together against the cold.

During the night, my camp-out was interrupted by only brief rain-squalls and wind batting at my tent — not a bear. By morning, the storm had dissipated and I returned to Plan A, which was to follow the railway over Cantwell to the purple mountainsides of Mount McKinley Park, above Healy and Ferry. There I let go of my railroad security blanket and launched off into the wilderness, taking a compass heading straight for Tanana.

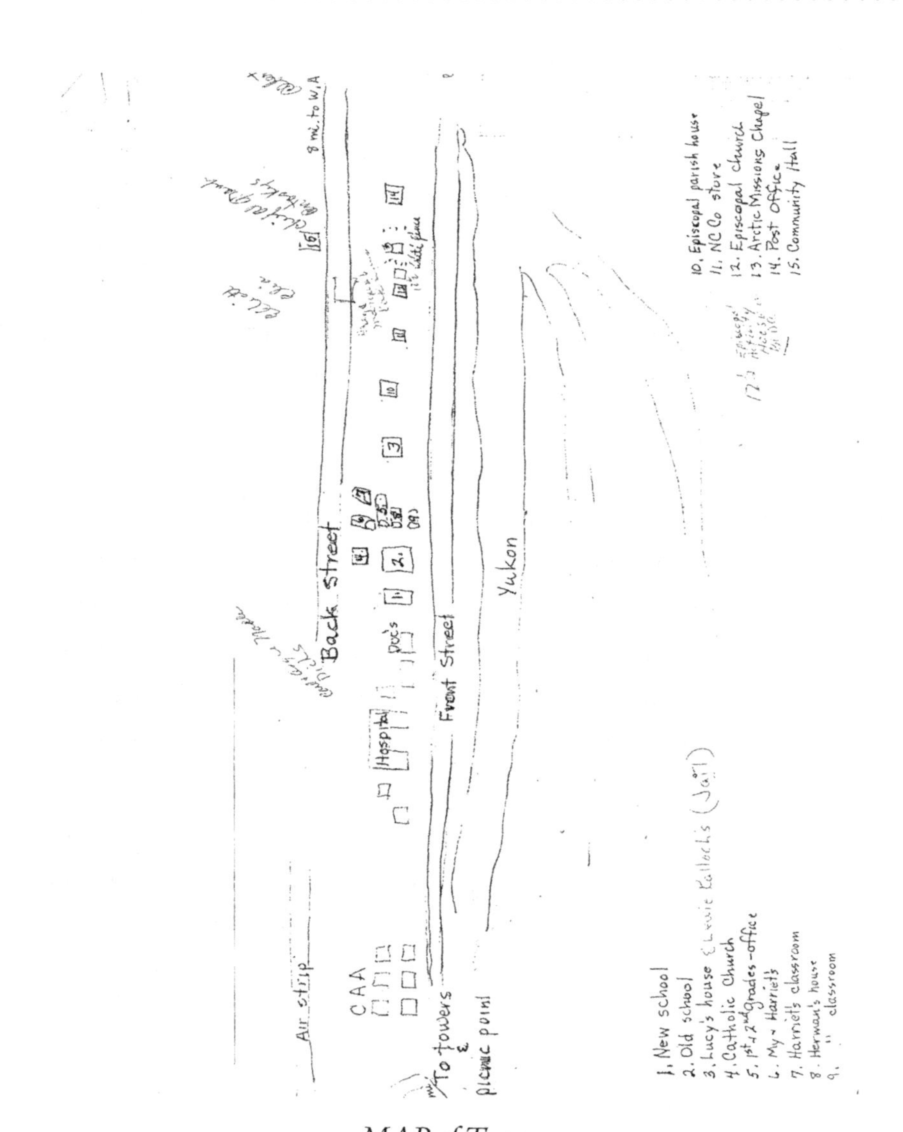

MAP of Tanana

(Courtesy of Anna Bortel Church)

Tanana, Alaska, 1957

In March, I'd visited Tanana to check out the 35-bed hospital and the village. Dr. Peter Hamill, the Medical Officer in Charge (M.O.C) took me around the attractive medical facilities, which included a white-with-red-trim, two-story, coal-burning hospital, adjacent nurse's quarters, and two new two-bedroom duplexes. The river, a mile across with an island in the middle, was an easy stone's throw from the duplexes.

Over the course of my two-week stay, I toured the village and met some of the people, including Dr. Hamill's wife, Margo, and their two children. The main road originated about seven miles west of the village at a cliff overlooking the winter-aged ice on the river. From there it traveled upriver and down a hill into the green-roofed, clean, white housing where the CAA employees lived near the airstrip, and toward the medical complex. The airstrip extended behind both the CAA area and the hospital.

Next came the white frame two-room schoolhouse, which in a previous life had been the Knights of Columbus building. The school teachers, the CAA personnel, some of the hospital staff, and missionaries were the primary non-Natives in the village.

Front Street proceeded past the school and along the front of a large two-story frame house, a remnant of the Fort Gibbon era when non-Natives built white two-story frame houses with big gardens of fruits and vegetables. I learned later that in the more recent past it had served as a jail, although at this time, Lewis and Lucy Kallock, a white man and Native woman, lived there. The street continued to Northern Commercial, the general store, where sporadically flown-in fruits and vegetables were anemic in color and exorbitant in price. The road dipped slightly and ran past the dark evergreen Episcopal Church. The Arctic Missions house-chapel followed next. (In 1988, Arctic Missions changed their name to InterAct Missions.) I'd knocked on the door and met the friendly missionaries, Roy and Marge Gronning. They were a study in contrasts — he was over six feet tall, and she not even five feet. Roy's short blond hair waved back over a high forehead and his blue eyes were pleasant and sincere. I knew instinctively that he'd be a friend.

Back Street, originating behind the schoolhouse, trailed east. One- and two-room cabins seemed tossed along the way. The sled dogs' houses, next to many of these cabins, served more as lookout stations for the dogs that seemed to prefer standing on top of these enclosures. After a while, Back Street disappeared into narrow dogsled trails among clumps of willows.

The road system primarily served the red Jeep-ambulance, several CAA vehicles, and the White Alice trucks. Since most people walked the short distance to wherever they needed to go, Ruby and I would be parting with our Chevy.

St. James Episcopal Church, March 1957

Tower House, March 1957

Tanana Chapel, March 1957

Tanana Post Office, March 1957

The main road continued for nearly eight more miles, past enormous gravel pits, a cemetery with occasional crosses poking through the snow near the old Mission site, and then into the hills to the "White Alice" station. "White Alice" was a United States Air Force telecommunication network constructed in Alaska during the Cold War "White" stood for the frozen north. "Alice" was an acronym for **Al**aska **I**ntegrated **C**ommunications & **E**lectronics. White Alice stations were operated until the late 1970s and replaced by satellite communication.

Tanana's history included another military defense. In 1988, Fort Gibbon was established to maintain telegraph service between Fairbanks and Nome. When Captain Charles Steward Farnsworth brought his wife and young son up north in 1989, he found a fort composed mostly of tents, incomplete barracks, and lean-tos. I couldn't believe anyone could spend an arctic winter in those conditions. Farnsworth rented a one-room log shack with a leaky sod roof. This inadequate barrack became the center of social activities for the enlisted men and upriver Natives, who would crowd into the room to eat dessert and listen to Helen Farnsworth play piano while her Black cook sang.

Until 1925, this outpost, along with four others in Alaska, maintained law and order during the gold rush.

Prior to this time, traders and Natives knew the location by the name of Nuchalawoyya (New-cha- la-WOY-uh) or, "place where the two rivers meet." Its central location contributed to active trading, as well as a place to discuss and settle problems. In 1869, the Alaska Commercial Company absorbed the Pioneer Company's trading post there and called the village Tanana Station. The Church of England had established a mission downriver, and in 1891, Reverend J.L. Provost, of the Episcopal Church, coordinated its relocation nearer the confluence of the rivers. A new mission was constructed, and in 1957, when Ruby and I moved to Tanana, it was hidden away outside of the village. The elegant structure of log and shake shingles was still an unusual and astonishing sight.

The Old Mission

At its peak, 512 people lived in the Fort Gibbon/Tanana locality, enjoying boardwalks, saloons, streetlights, and other amenities. Hay and potato fields spread behind the small city which boasted the tallest building in Alaska — a four-story hotel with a cupola. After World War I, however, most of the facilities were demolished. With lack of military incentive and the decline of gold seekers, there was not much to sustain the settlement.

Now, in the 1950s, there remained few vestiges of such bursting activity. What wasn't destroyed was taken back and concealed by the voracious and insidious undergrowth. The population had declined to just under 300 Natives and non-Natives.

Tanana *did* retain one significant distinction, and it was for that reason I was coming to this area: medical care. The first hospital had been built in conjunction with the Episcopal Mission and established about 1908. Its easy access along the Yukon River, the presence of military physicians, and the fact it was the nearest medical services for many of the Interior Alaska villages made Tanana a medical center. When the hospital burned, it was replaced by St. Agnes Hospital in 1914. The Bureau of Indian Affairs assumed responsibility of the hospital in the 1920s. But I was to be in the new hospital, constructed in 1949, and transferred to Public Health Services in 1950. The first PHS physicians had been Dr. Andy Wehler (1952), Dr. Jean Persons (1953-1956), and then Dr. Peter Hamill. It was an integral part of Interior healthcare, and, in addition, played a major part in the village economics, providing work for numerous people.

Decades later, it would become more efficient and economical to transport patients to Fairbanks or Anchorage hospitals. By 1980, the Tanana hospital would average a daily patient occupancy of five, and on October 1, 1982, it would be abandoned. A health center would be established. In 2009, the hospital would be demolished with a closing ceremony for the villagers and previous physicians and employees on September 10, 2009.

I'd flown with another bush pilot on my first trip to Tanana. Now I was on my own. Before I had left Anchorage, a CAA weather briefing advised me of forest fires, started by lightning during the extremely dry season. I expected to see billowing black smoke or yellow-orange flashes of fire. I saw neither, but I smelled the smoke. Even though I knew of the forest fires, my immediate reaction to the acrid odor was fear — fear of fire in the plane. Startled, I looked around the cabin. Nothing aflame. Then I looked outside again. A strange haze was moving toward me.

The vapor was not just a cloudy weather condition — it was actually smoke from the forest fires. Although it was about 10 a.m., the sky wrapped me in an eerie gray-brown veil. At first, the sun appeared haloed in a brown shiny ring, but then, as if controlled by a dimmer switch, the sun gradually faded from sight. If it weren't for the sensation of dry heat, the smoke could have been mistaken for damp fog. Forward visibility deteriorated rapidly from five miles to two miles. I dipped a wing and looked toward the ground where visibility remained.

Landmarks disappeared. Rivers and lakes below all looked alike. Low spruce-covered hills rolled and repeated themselves beneath me. I wanted to trust my compass heading, but something seemed wrong. Later I learned that this area was noted for severe magnetic variations. The smoke crowded around me until I had only one-mile forward visibility. An hour passed. Somewhere I'd missed my main checkpoint, the Kantishna (Can-TISH-nah) River. Where was I?

As if these frustrations and anxieties weren't enough, my gas tank bobber wasn't moving. I'd learned to refill the gas tank from the cockpit by inserting a rubber hose into the five-gallon gas cans in my baggage compartment and pumping out the gas into the nose tank with my wobble pump. This procedure required some tricky maneuvering within the cockpit; and more so when I was trying to keep some kind of bearings. But I did it.

With that accomplished, I needed to get on course to Tanana. At this point, Plan C was Common Sense, not Compass. I chose the first small stream and followed it downstream. Eventually all running water in this area would wind up in the Tanana River. Even in all this smoke the Tanana River could not be missed; after all, it *was* over 600 miles long and averaged 200 to 400 feet wide, with many side channels and sloughs. Since it was the Yukon's largest tributary, it would surely lead me to my destination. Despite my confusion, I felt as though I was finally getting somewhere.

After a while, a large river came into view and as I followed it for several miles, I was able to identify Manley Hot Springs. Apparently, I'd followed the Zitziana River to the Tanana River. I straightened my tired neck and stretched my shoulders, relieved that I was on the right track, *and* that I had a potential emergency landing strip beneath me.

Before long, I came to a wide shallow area, which I learned later was called Squaw Crossing. On this flight, there was no significant disturbance, however, in the future, I came to respect the area. It had an ominous reputation for severe turbulence from the strong winds coming over the hills to the north, from the Yukon River.

About an hour out of Manley Hot Springs, I crossed off my last checkpoint, the confluence of the Tanana and Yukon rivers. Right on target and two miles past this point, I spotted Tanana — a beautiful sight for smoke-filled eyes.

After circling the village and checking out the river along the shore in front of the hospital, I splashed down on the legendary Yukon River and taxied against the current. Several people ran down a wide trough in the bank to meet me. Before I really let out my tightened breath, I wanted to secure the J-3 to the narrow beach shore. After some maneuvering on the swift-flowing river, I had it safely tethered. I'd made it!

Now that I was here, I looked forward to starting my official duties. Well, maybe not right this minute. I was glad it was Saturday night with Sunday coming. I could use a day of rest before getting into the swing of things.

Honoring our Sacred Healing Place, Tanana, Alaska

Naomi Gaede-Penner was asked to edit a publication for the 2009 commemoration ceremony. The booklet, "Honoring our Sacred Healing Place, Tanana, Alaska," documents the history and cultural impact of healthcare in Tanana. Numerous photos, lists of hospital personnel, and stories from people whose lives were touched by the hospital, illustrate the significance of the Tanana Hospital Complex. A pdf of this publication can be found at www.prescriptionforadventure.com.

CHAPTER 8

A STRANGE VILLAGE WELCOME

July 1957

Elmer at the Tanana Hospital, 1957

HERE AT THE TANANA HOSPITAL, sickness seemed out of place. What a contrast to the 400-bed Anchorage hospital filled with long-term tuberculosis patients. Summer saturated the air and permeated every corner of the building. I'd arrived in the early afternoon, and with no time to spare, had been re-introduced to the hospital staff and oriented to the building. Now I was the M.O.C. Besides the responsibilities of sole physician, I was the administrator of 38 employees. My geographical oversight would encompass a region the size of Kansas. Every morning and evening, school teachers and missionaries in the villages would expect me to contact them through the hospital radio and discuss medical problems.

Now, even though it was early evening, the Alaska sun had no intentions of resting for the night. In bright daylight, I leisurely walked through the vacant waiting room onto the hospital's front porch. Hanging pots of red and white petunias flounced in the gentle breeze and hosted a friendly greeting to any prospective patients.

I looked out on Front Street where several children stirred up the powdery surface with their bicycles. Beyond them the Yukon River hurried toward the sea. A "kicker" boat (what the Natives called an outboard motor) fought the current and buzzed loudly up the river. This was the idyllic start to my assignment in Tanana. It didn't last.

"Doctor! Doctor!" The cacophony of words and door-pounding finally turned into intelligible sounds. I rolled over in bed and looked at the clock: 5 a.m. This was Sunday, the morning after I arrived in Tanana.

What do they want? I asked myself, climbing out of bed. *And who were they?* My uneasiness gave way to curiosity and I barefooted to the backdoor. I leaned my face near the door and cleared my throat.

"This is the doctor. What do you want?" I asked, trying to sound calm.

"Lee shot Floyd," a voice stood out amidst the clamor.

And that was my welcome to Tanana. I'd thought my role would be medical administration and physician, not Justice of the Peace. In the future, I'd learn how to go about dealing with such matters and I'd rely on the village chief for some directions. But right now, I felt caught off guard, baffled, and unprepared.

Later, I was apprehensive about telling Ruby about this traumatic event and wondered if she'd second-guess our decision to move here. To my relief, when she did arrive with the children several weeks later, she took it in stride and plunged into village life. She was a suitable partner with her hardiness, management skills, and ability to do what it took to establish home and hearth for our family.

Her imminent and ongoing undertaking was to keep Mark from toddling into the river. Even though Front Street was right outside our house, "playing in the street" posed no great danger since there was only occasional vehicle traffic. And, if he would have just *stayed* in the street, that would have been desirable; but the steep river bank was only a short distance farther. He was a busy child, not one to be confined, prone to exploring the insides of gas ovens, clothes dryers, toilets, and the great outdoors. To his safe-keeping benefit, Ruth tended him like her own child — even when he drove her to tears by throwing her kitten into the dryer. Often, she pulled him in a wagon or on the back of her tricycle. Surprisingly, during our time in Tanana, none of the village children ever fell into the river; certainly, a miracle, given its proximity.

Ruby's next task was to get the girls into school. Naomi and Ruth joined the 51 students at the first through eighth grade Tanana Day School. Anna Bortel, a young vivacious teacher from Ohio who had arrived shortly before we did, taught the first and second-graders on one side of the double-doors that separated the open space into two classrooms. Florence Felkirchner, a staid, older woman who had taught in Alaska for some time, taught grades three through eight. For high school, students were sent to boarding school at Mount Edgecumbe in Sitka or Wrangell Institute in Wrangell.

Naomi, a second-grader, liked Miss Bortel and made friends fairly quickly. Ruth's nature was to stand back, watch for a while, and wait for an invitation to join in. Nevertheless, it wasn't long before both girls adjusted to life in the village. Naomi's closest friend was an Athabascan girl, Sally Woods. Naomi was intrigued by her earrings of tiny dried spruce needles. Actually, Sally's ears had been pierced with a sewing needle. Dental floss dipped in iodine held the opening from growing shut; however, when her ears became infected her mother used spruce needles to offset the infection. Naomi didn't ask for earrings, but she did let her hair grow into long braids — as did Ruth.

At our house, the girls baked oatmeal cookies, played with dolls, and made mud pies. Then, they'd go to Sally's cabin at the other end of the village. Sally's mother sliced fish adeptly, hanging them on drying racks made of spruce poles, unprotected from the rain, ravens, and flies. The recently caught fish shone a bright oily red, while the older fish were dark with slightly curled edges. These fish provided sustenance for both the people and their dogs.

Ruby had been forewarned that groceries would be more limited here than in Anchorage. That was true. Even after the stern-wheeler churned down the river to deliver the years' worth we'd ordered, food preparation was restricted by the lack of fresh fruits, vegetables, dairy products, and baked goods. This was perplexing for a farm girl who had grown up with such items within an arm's reach.

Yukon barge, July 1958

Consequently, Ruby baked nearly every day and became quite proficient at shaping hot dog and hamburger buns for summer picnics. With culinary ingenuity, she took powdered items and wild meat and presented us with bon appétit delights. I supplemented her limited resources when I made infrequent medical trips to Fairbanks. Before I'd leave, the family gathered around and helped scribble out a grocery list. Ruby craved fried potatoes. I brought her potatoes and cottage cheese. Ruth begged for bananas. Naomi wanted store-bought white bread. Mark had no preferences. I bought myself a ring of bologna.

Clothes shopping posed another challenge. We ordered by mail from the Sears Catalog, which took several weeks for a package to arrive. We were never sure what we would get. Sometimes items would be substituted or the clothes represented in the pictures didn't look the same in real life. Shoe purchasing consisted in choosing a shoe style, tracing around each foot, and mailing this information to Sears. Later, Aldens and Montgomery Ward catalogs offered other mail-order options.

Articles such as sewing machine needles, clothes patterns, glue, gift wrap, and birthday cards were purchased on seldom-made trips to Anchorage by Ruby and the children.

Ruby's social life helped her manage the isolation, especially when winter set in. We were farther north now and the sun had more difficulty staggering above the horizon than in Anchorage. Ruby had regular game nights at our house and a sewing group, which included nurses, CAA personnel, Anna Bortel, and White Alice workers' wives. In addition, our house became a revolving door for missionaries isolated along the river — as if Tanana wasn't itself a Bush village.

I'd wanted adventure, but that first morning when I woke up to door pounding and a murder announcement, that concept was defined in ways I hadn't anticipated.

"Floyd's friends want to shoot Lee," another man's voice added to the previous headliner.

Shaking off the daze of sleep, I realized that my life was not in jeopardy and I opened the door. Three Native men were crowded onto the narrow concrete back steps, their brown eyes were large circles. I caught the smell of alcohol on their breath. They pressed me for a plan to prevent the lynching.

I stalled, "Let me get dressed. Then take me to your chief."

After only a few moments, they impatiently banged on the door and urged me to hurry. This was not going to be how I imagined meeting the chief. When we arrived at his cabin, the young Alfred Grant introduced himself. He was not a strapping man as one might expect, but thin and unassuming. With no social chitchat, we got right to the point. Together we decided that the only way to prevent further violence was to hold the murderer in a cabin guarded by four men.

The chief, and the men who had escorted me, left the cabin to implement the plan. I returned to the hospital and with the two-way radio summoned the Alaska Territorial Police in Fairbanks for assistance.

At that point, every person was a stranger to me, as well as their culture, civic process, and village social norms. It wouldn't be until Ruby and the children arrived that I'd feel more at home and we'd integrate into the village and join in community events. One of the community gatherings was the potlatch which marked events such as births or deaths. This included a feast, story-telling, dancing, and often, gifts.

As were most village events, potlatches were held at the Community Hall. We weren't sure what to expect our first time, but were more excited than anxious. Benches lined the perimeter of the community hall and two barrel stoves squatted on the recently scrubbed splintery board floor. Villagers who arrived at 6 p.m., seated themselves on the benches. Five-foot-wide oilcloth runners were spread before these benches and the latecomers then sat on the opposite edges facing the benches. Like everyone else, we brought our own bowls and spoons for the soup, which was ladled from a sawed-in-half five-gallon Blazo gas can and served by men walking down the center of our oilcloth "table." The soup's content varied, depending on the donated meat and other items. We could expect a mix of canned vegetables and macaroni — and not much salt.

After the soup, we were served meat — usually moose and sometimes caribou. Depending on the physical size and age of the individual being served, the servers cut off custom-size chunks. In most cases, the fist-size portion was eaten with the dexterous combination of fingers and a knife. Very carefully, the meat was bitten with the teeth and then the knife was used to cut it off dangerously near one's lips. Amazingly, I never had to suture anyone's face as a result of this method.

Our meal progressed to Sailor Boy Pilot Bread, a thick, three-inch round, salt-less cracker, which was a staple of every household in Alaska. Butter, strong hot tea, canned peaches, or fruit cocktail completed the meal; and then the children were given gum from the general store and the adults, cigarettes.

We appreciated the Native's acceptance and these were good times to mingle.

Later on, the Spring Carnival in April provided another opportunity for our family to learn about village customs. School was dismissed and Natives from surrounding villages joined in the celebration. The frozen river furnished a slushy arena for the dogsled and snowshoe races. There were categories for men, women, and children. The women often did better than the men since they didn't drink as much alcohol.

The riverbank hot dog stand seemed incongruent and more appropriate for a baseball game, but the Native man who managed it did a booming business. Ruby and Anna Bortel, the girls' school teacher, took turns volunteering to sell refreshments at this main attraction.

Ruby and I also participated in the Fourth of July festivities which included sack races, bike races, egg-carrying races, and field events. Ruby ran with an egg on a spoon from the general store to the Episcopal parish hall, came in second, and won $3.00. I surprised the villagers by placing first in the broad jump, winning $5.00. I overheard them muttering among themselves that they didn't think a soft white man could compete.

What they didn't know was that my brother, Harold, and I had run track. Now, if they wanted to see more competition, they should have seen our farm-boy basketball team take on the city boys. All those perspiringly hot days of bucking hay bales, shoveling grain, and cleaning the barn hadn't been wasted.

All this village familiarity and fun, however, was unknown to me on that first day. Instead, I was immensely relieved when my first unofficial day at work came to an end and the Alaska Territorial Police showed up. In a strange parade, Chief Alfred Grant, the murderer, and a group of Natives proceeded to the airstrip with the dead man in the back of the old black hospital pick-up truck. When the plane took off and headed back to Fairbanks, the murderer found himself with an odd seating companion — his victim. *That* had been another oddness of the day. It was the custom in this territory to sit the corpse in a chair. Rigor mortis would then shape the body to more easily fit in an airplane seat, often beside the pilot.

A lot of strangeness had happened and I hoped my initiation was completed. By any means, it was a clue as to what life as a village doctor could be like. Time would tell if I was up for the challenge.

CHAPTER 9

HOUSE CALLS, ALASKA STYLE

February 1958

THE WINTER OF 1958 would not release its hold and for weeks the arctic temperatures hovered between 40° and 50° below zero. The cold crept under window sills and around door jams, following and finding every living thing. On rare, clear days, the darkness would lift for several hours, allowing us freedom from our winter box. Not many people ventured outside, and the silence accompanying the dark cold was broken only by sled dogs howling in chorus atop their houses, or by the whine of a chain saw finding fuel to fight off the cold.

On this particular day, I maintained my normal routine, and by noon I'd examined a woman with tuberculosis, children with draining ears, and a large number of people with respiratory infections. Taking a break from face-to-face patient contact, I made my daily radio contact with the villages both up and down the Yukon River. It wasn't long after calling in that I realized this was not to be a usual day.

My first call connected me with Manley Hot Springs, a small village upriver, along the Tanana River, about 90 miles west of Fairbanks.

"Doc, Charlie's awfully sick." The Manley Hot Springs' innkeeper's voice crackled on the other end of the two-way radio. "He didn't show up for several days, so we checked up on him. Seems he's had stomach problems in the past, and it must be acting up 'cause there's blood everywhere. I mean, Doc, it's on his clothes, his bed, the floor... Doc?"

In Manley Hot Springs, Gil Monroe, the innkeeper, was the only person with a two-way radio. He was in charge of out-of-village communication and for arranging charter air service when necessary.

"Sounds like a bleeding stomach ulcer." I tried to sound matter-of-fact as I mentally considered possible medical treatment.

"I'm afraid we're going to lose him." Gil's anxiety mounted.

"Try to give him some broth, tea, Jell-O, or powdered milk." I wanted to give the innkeeper some sense of control — and hope, but felt helpless at this distance.

"Doc, I wish you could come," he pleaded; then added bleakly, "The weather's bad and nothin's flying."

Here in Tanana the ice fog had dissipated and, in its place, the warmer temperatures of minus10° brought snow squalls.

"Why don't you contact Fairbanks and request a plane for medical evacuation," I advised. Believing full well that the innkeeper was not minimizing the crisis, yet uncertain that assistance from Fairbanks was realistic, I added, "Meanwhile, I'll try to crawl under these squalls and get in."

I was caught between the need to attend this critically ill man and to stay at the Tanana hospital, where I expected a young woman to go into labor at any time. I discussed the situation with the nurses. They assured me that they could take care of the delivery, if necessary,

Without a doubt, these bush nurses could handle any situation; still I felt apprehensive as I bundled up to check my plane. I tried to keep my plane in readiness for these kinds of emergencies. A 150-foot extension cord trailed from the house, across the road, and down the 12-foot bank to the J-3 on skis. This umbilical cord supported a 200-watt bulb tucked into the plane engine, which in turn was wrapped in blankets and canvas. At these immobilizing temperatures, my plane was ready to go.

It was after 12:30 p.m. I had two more hours of pallid daylight. I only needed 45 minutes of flying time, yet I moved quickly. My emergency medical supplies were packed and waiting for these kinds of mercy flights, along with a sleeping bag, axe, gun, engine cover, one-burner Coleman stove with four-inch stove pipe to funnel heat under the cowling, and a small can of stew. I calculated every ounce when flying. If it wasn't a life-saving essential it was left behind.

The CAA weather report indicated marginal flying conditions, and I acknowledged that setting down unexpectedly en route was highly probable. In this unforgiving weather I could leave nothing to chance or make any mistakes.

After a routine preflight check, I took off on the cleared, but rough, river ice. I climbed to 500 feet and followed the Yukon River for a mile, then continued up the Tanana River. There was a moderate chop over Squaw Crossing where the wind funneled through a saddle in the hills to the north. Visibility was poor but adequate as I clung to the north bank of the river.

The tension in my neck eased as landmarks indicated that I was only 15 minutes out of Manley. Then, without warning, oil splattered up over my windshield! I glanced down at the oil pressure and engine temperature. Normal. What was wrong? Oil continued to ooze out of the right cowling and within minutes my forward visibility was zero.

Out my side windows, I searched for an emergency landing spot on the river. Soft new snow deceptively smoothed over the treacherous pressure heaves formed by the tug-of-war between churning water and restricting ice. When the water would win, huge sheets of ice would break and push diagonally above the other river ice. Only an occasional jagged edge, however, poked through the snow, hinting of the larger danger.

Aware of the odds, I selected a reasonably smooth area, reduced the power, and tried for the gentlest landing possible. The plane bounced against the chunks of ice and finally jolted to a stop.

Cautiously, I crawled out of the plane. The skis were still firmly attached to the aircraft — a positive sign. I loosened the cowling, trying to avoid the sticky, black oil. What a dirty sight. But more good news: three quarts of oil remained — enough to make it to my destination. Then I found the culprit, a dislodged oil cap. The small offender sure had created a big mess — and hazard. I screwed it on easily.

Cleaning the windshield was not so simple. The freezing temperatures added to the difficulty. I scraped the now snow-glazed molasses-like oil from the windshield and then used snow to finish the cleaning process. Looking up from my project, I noticed that visibility was down to a mile now. I needed to get back in the air. Scouting ahead on the river, I checked for the least hazardous takeoff path. The J-3 did its best, given the conditions, and I resumed flight to Manley.

Manley had a flow of hot water. Because the location was easily accessible by water, a trading post was established in 1881. In 1901, a prospector, J.F. Karshner homesteaded the spring site. Thousands of expectant miners flooded the area when other prospectors struck gold in the nearby Eureka and Tofty areas. At the same time, the U.S. Army set up a telegraph station.

One of these men, Frank Manley, differed from the empty-pocketed others: already he had several hundred thousand dollars. Consequently, Manley's money and Karshner's hot springs joined hands to start Alaska's first substantial geothermal resource project. This fortuitous relationship encountered one snag when word got out that Manley's real name was William Beaumont. Accused of horse thievery, he'd left Texas and assumed an alias. Sent back to his homeland, he was acquitted and returned to Manley.

Beaumont, alias Manley, and Karshner cleared land for farming and built a 60-room hotel, along with other buildings for raising poultry, hogs, and dairy cattle. In 1910, the future held promise and "Hot Springs" had a booming population of 101 people. In this era, steamers brought guests up the Tanana River in summer, and in winter a two-day overland stage-coach transported people from Fairbanks. Mining, however, declined, and then in 1913, the hotel burned, ending this glorious era of history. By 1930, only 45 residents remained. In 1957, "Hot Springs" was changed to Manley Hot Springs. Now, in 1958, the village of half Natives and half Whites, vacillated between 20 and 30 people, depending on the hunting and fishing seasons.

(In 1959, completion of the Elliot Highway would provide a summer road-link with Fairbanks, and in 1982, the highway would be maintained for year-around use.)

"Doc, so glad you make it!" Gil greeted me after my plane engine shuttered into silence. He was a non-Native man between 50 and 55 years old. His army-green parka with fox fur ruff had snow accumulating on the shoulders. "Ol' Charlie isn't doing well at all. He can't eat, can't get out of bed... boy, am I glad to see you."

Snowflakes clumped together and began to blanket the plane. Gil tied down the plane as I drained the remaining oil into a two-gallon can. There was no cord here to sustain life while the plane waited for my return and I'd need to warm the oil before I took off later. Carrying the can with me, I tried to find out more about Charlie Stark from the innkeeper.

"Well, Doc, he's one of the ol' timers. Been in Alaska since somewhere about 1880 — you know, when all the others came up to the Last Frontier with gold fever. He's an independent and tough old codger — a survivor. Worked on the road from Fairbanks to Rampart, mined gold around the Eureka area, and of course homesteaded, too.

Manley Hot Springs Lodge/Roadhouse, February 1958

We reached the innkeeper's two-story roadhouse, which provided eating facilities, a lounge, general store, and upstairs sleeping accommodations. A shed housing the generator joined the few houses clustered around the roadhouse. Several hundred yards farther, I could see the steam rising from the hot springs. The commissioner's log cabin office and a small bath house nestled near the warmth.

My bunny boots alternately slid and crunched on the hard-packed snow as I tried to keep pace with Gil. Charlie sure was lucky to have a friend like him.

"Yep, that ol' fellow's at least 85 years old. Been here in Manley Hot Springs for about15 or so years — everyone likes him... Poor guy, doesn't seem to have any family, though."

Gil knew a lot about Charlie, probably a lot about everyone in Manley Hot Springs.

"That's Charlie's place. You should see his garden in the summer. Big ol' cabbages, plenty of carrots, and piles of potatoes."

It appeared that Charlie was upholding the Manley Hot Springs' legend, which began in 1910 when 150 tons of potatoes were shipped downriver to the Iditarod (Eye-DIT-uh-rod) mining district.

"...things wouldn't be the same without Charlie around."

Yes, Charlie sounded like a lively character, although right now there didn't seem to be much life in the cabin in front of us. The tin-roofed structure leaned to one side as if the snow piled against it was trying to push it over. The windows stared back at us blankly, and only a thin thread of smoke came from a straight stovepipe, the vapor blending into the early afternoon dusky sky. At 2 p.m. the sun was going down.

We climbed up the several steps to his porch and stomped the snow off our boots — a kind of notice that we had arrived.

Gil tugged open the door, and we walked into a dark cold room. The paling outdoor light did not penetrate the interior and only a dim kerosene lamp flickered on the faded oilcloth-covered table. Three heavy wooden chairs grouped around this small, sturdy table, and wooden shelves nearby displayed a meager assortment of canned goods, pancake mix, a round box of oatmeal, and a half-empty bottle of syrup. Old carpeting and furs lined the rough planked floor. Slowly my eyes adjusted to the darkness, and I set the two-gallon can of plane oil on the flat top of the barrel stove. As I approached a low metal army bunk bed against one wall, the putrid smell of vomit confronted me. There was Charlie. Our eyes met in the darkness.

"Charlie, I'm Doctor Gaede from Tanana. I came to help you."

Charlie, weak but conscious, tried to nod.

Gil pumped up the kerosene lamp, which slightly brightened the room.

I could see dried vomit on Charlie's flannel shirt, army blanket, and the floor around his bed.

"Charlie, you're looking pretty rough. Tell me what happened."

"Well, Doc," he began feebly, pausing often as he spoke, "I've had stomach trouble before and had some bleeding a couple years ago... I saw a doc in Fairbanks and he gave me some medicine, which cleared it up...now lately I've had stomach pains again and not much appetite...the past couple days I've been throwing up everything I eat."

Gently pushing his blankets to one side, I began my examination. The shriveled skin of his large-boned arm folded into the blood pressure cuff. His blood pressure was down to 80/40 with a faint rapid pulse of 120. Carefully, I rearranged his flannel shirt, exposing skin drawn tightly over prominent ribs and a slightly bloated stomach. Yes, his stomach was tender... liver barely perceptible...heart and lungs okay. He was larger than I first thought, probably six feet, but only about 150 pounds.

My eyes had adjusted to the dim room, and focusing back on his face, I saw stubs of teeth, coated with dark, dried blood, contrasting with ghost-white skin. The diagnosis was obvious: hemorrhaging gastric ulcer.

"All right, Charlie, let's see what we can do to make you more comfortable."

Our work was cut out for us.

While I had been examining Charlie, Gil had stoked the fire in the barrel stove and started heating water. Now he began cleaning up the cabin, locating clean bedding, clothes, and towels. In the process, he found some canned soup and warmed that on the stove.

Meanwhile, I started an IV and gave him the requested "pain shot."

The innkeeper returned to his inn and I fed Charlie spoonfuls of soup broth. In a short while, Gil came back and announced, "I got through to Fairbanks, but they're socked in, and planes aren't moving. You won't get any help from them tonight, or until the ice fog lifts."

Sensing my concern, Charlie tried to reassure me, "It's okay Doc, I've lived a full life. I'm ready to meet my Maker."

If I could just pull Charlie through tonight, he'd be all right, but it would be a long night.

After giving Charlie two bottles of IV, he became more comfortable and his vital signs stabilized. The three of us began exchanging Alaska stories of our experiences. Eventually we all fell asleep; Gil and me in makeshift beds.

Suddenly, I was awake. Something was wrong. Like a mother sensing trouble with a child, I got up to check Charlie. The flickering kerosene lamp cast gentle shadows on his peaceful face. I checked his blood pressure and pulse. Charlie was going into shock.

I'd already used up my two bottles of IV solution. Now I'd have to improvise. The water in the bucket on the stove was still warm, and at one time had been to a boil. Hurriedly I filled an empty IV bottle with the water, added one teaspoon of salt from the lone salt shaker on the table, and dashed out the door to cool the saline solution in the below-zero snow. Moments later, both my wool-socked feet and the homemade solution were lukewarm. I returned to my patient, quickly found a vein, and started the IV.

I waited. Slowly the bottle emptied. In the same fashion as the first, I prepared a second. Finally, Charlie's blood pressure was perceptible and I slowed down the IV drip. Minutes stretched into hours as I kept the night watch. The innkeeper snored unperturbed.

About 10 a.m., dawn pushed its way through the frigid darkness and into the cabin and all three of us awoke. Returning from a second call to Fairbanks, Charlie's "guardian angel," grin gave away the good news.

"The weather's breaking! Chances are we'll have a plane by noon!"

Charlie tenaciously clung to life as he waited for the rescue plane and then endured the 45-minute flight to Fairbanks.

Later I learned that once he arrived at the hospital, he received a walloping eight units in blood transfusions.

As Charlie's plane took off, I put a hand on Gil's shoulder. "Thanks for standing by me."

For a moment he didn't say anything; he just swallowed hard.

"Yeah, Doc," he answered gruffly. "Up here we have to stick together and make do with what we have."

The engine oil was warm from its resting place on the stove, and the plane fired up easily. The small craft seemed eager to return home and put on its best behavior. No unexpected surprises this time. Fresh snow covered the river with no hints of the previous emergency landing. In contrast, I knew that the crisis with Charlie would leave its mark on me. I thanked God, who I acknowledged as the Great Physician, for the opportunities and abilities He had blessed me with — and flew back to check on the pregnant woman.

You see, this was no ordinary pregnant woman. This was an Inupiat (In-YOU-pee-at) Eskimo woman offering her child for adoption — and Ruby and I were to be the parents of the gift. I was both disappointed and elated to discover that in my absence, a baby girl, our Mishal (Mih-SHELL), had entered the world and our lives.

And Charlie? Even though he may have figured his full life was complete, his body fought back, granting him several more years at Manley Hot Springs.

Manley Hot Springs Roadhouse

The Manley Roadhouse has received customers nearly continuously since 1903, and is the oldest functioning roadhouse in Alaska. For more information or reservations call the roadhouse at (907) 672-3161. https://www.manleylodge.com.

CHAPTER 10

A SIMPLE CARIBOU HUNT

February 1958

"YOUR SKI LOOKS FUNNY."

Alfred Grant, now the *previous* chief of Tanana, nudged me from the backseat of the J-3. We were trying to go caribou hunting. Alfred needed meat and also wanted to check his beaver trap-line near the caribou feeding grounds. I was eager to fit into the community and to come alongside the Natives; and besides, I'd never shot a caribou. When the young Athabascan sought me out, I jumped at the chance.

This was our second attempt — within the same morning. In mid-January, darkness invaded most of every 24 hours and "morning" lasted about two hours. On cloudy days, the sun played hooky completely. But to our advantage, this Saturday, in a rare display of benevolence, the sun had proved it had not deserted us entirely. The winter world was smothered with snow, but not as cold as usual; although minus 13° was still bitter enough to pinch our nostrils and frost the edges of our parka ruffs. Wind added its chill.

Alfred pointed to the left ski. There it was, hanging forward with the front tip downward.

"For cryin' out loud," I muttered. "I just had this thing in Fairbanks for its annual inspection, and ..."

We were a short distance out of Tanana with a destination of the northern tributaries of the Kuskokwim (CUSS-do-kwim) River, about 75 to 90 miles south of Tanana. A large caribou herd usually foraged there, so it would be like walking up to a meat counter. Simple.

I assessed the problem. "Looks like a broken cable," I shouted over my shoulder.

We'd already aborted this trip earlier. Subsequent to our initial takeoff, I'd noticed the airspeed indicator was not representing the true speed. Just as I needed to know human anatomy to practice medicine, as a pilot, I needed to know airplane anatomy. Not that the J-3 really had much anatomy beneath its thin skin. My in-air diagnosis was that we'd lashed our snowshoes too closely to the wing strut pitot tube opening, which measured velocity at that certain point. The problem was annoying, but easily remedied. We had landed and retied the snowshoes in an unobstructing position. Time was ticking on the short day. We needed to get going before the shadows too quickly lengthened and the early night pulled down the thermometer.

Following the next takeoff, I scanned my instrument panel and everything seemed to be functioning correctly. The wind increased and with it, turbulence. Bouncing around didn't bother me, or my passenger, but apparently it didn't agree with the Piper. About five miles out of Tanana I abruptly had difficulty controlling the aircraft. That was when Alfred had anxiously tapped me on the shoulder. I didn't have to tell him that we were in serious trouble. The crippled plane was not going to make landing easy.

I turned back toward Tanana. A strong headwind aggravated the already limping plane. I fought the controls. Alfred sat stoically in the backseat, not saying a word, or if so, I couldn't hear it.

Since the J-3 did not have a radio, I couldn't transmit my plight to the CAA. I needed sky-writing. Around and around, I patterned above Tanana, hoping someone could hear the plane's repeated buzz amidst the bluster. But more critical than that, it was imperative they discerned my dilemma and could provide ground assistance.

After being buffeted around for more time than I cared, I saw several men run out of the CAA station. I strained my eyes to figure out their strategy. They looked up at me with hand-held fire extinguishers. They planned to put out a fire. This reality spiked my blood pressure — and consoled me all at once. The emergency crew made its way toward the middle of the runway. I wasn't sure what else they could do. Bringing out the ambulance would have alarmed me more. *I* was the doctor, and supposed to be the one saving lives in such a crisis.

During this time, Ruby heard aircraft noises. Airplanes have their own distinguishing reverberations, and she'd learned to recognize the J-3, but perhaps the background wind distorted the familiar pitch. Within this context, she assumed the airport was unusually busy with a number of bush and charter pilots.

My plan was to land on my right ski, which would cost me my right wing, but not our lives.

Unknown to all of us, there was another plan. Anna Bortel, the girls' school teacher, heard and recognized my plane. She saw the dangling ski and knew that most certainly I would crash on landing. Terrified by this impending tragedy, but not knowing what to do, she dropped to her knees and prayed frantically and fervently that God would stop the wind and somehow save my life.

Meanwhile, I swung around on base-leg and started my descent. The plane bucked in the unstable air. The snow-packed airstrip moved toward me rapidly.

"Hang on!" I yelled.

This would be easier for Alfred than for me. I was going to fly this thing literally into the ground.

My hand gripped the stick tensely. Just before touching down, I jerked it back to try to swing the left ski tip forward before stalling onto my right ski. There was no time for a counter move. I wouldn't be able to actually see if this worked, we'd know it when we hit — or ground-looped. A split-second later I felt the aircraft settle onto both skis equally — on the ground. The renegade ski must have moved into position. We continued in a straight line. Like a wind-up toy losing power and ending its energetic cycle, we came to a rest in front of the CAA delegation — all armed for disaster and in position with their feeble fire extinguishers.

Their anxious faces eased into grins when their eyes met mine.

I sat stunned and unmoving. We'd actually made it down — in one piece —and not in a million particles littered all over the airfield.

I finally came to my senses and shut down the engine. Then, I turned around to look at my passenger. If a brown face could be white, his was, and expressionless. For the first time since he'd made the near-fatal announcement, Alfred spoke up, "Good job, Doc."

When I opened the plane door the men clustered around. "Boy, were you lucky, Doc!"

I didn't exactly agree about the luck; I recognized a miracle when I saw one. God had quieted the winds. At the higher altitude, the gusts had pushed back the weakly cabled ski, but then near ground-level had subsided, allowing the ski to become horizontal.

"Alfred, do you still want to get that caribou?" I figured there was no way he'd get back in that plane.

He looked at me for a moment. "The sun is still here."

I took that as an affirmative.

Everyone lent a hand in repairing the faulty cable, and sure enough, without a word Alfred climbed back in and fastened his seatbelt.

Once we were airborne, I'd catch him looking out either side and checking on the skis. Every now and then my co-pilot would inform me of their status.

"Ski looks okay." "Tip not down." "No problem." "Ski good."

There was no need to get too high in the sky and besides, the lower altitude afforded us a wildlife tour. Browsing moose showed up most frequently, but then Alfred pointed out two loping wolves. After 40 miles, I noticed numerous tracks on every lake we flew over.

"The herd must be here somewhere," I told Alfred.

We scouted the hilly country but only found more moose. I continued farther south until we came to the north fork of the Kuskokwim River. Scrutinizing the countryside, I caught sight of two caribou below my right wing tip. Alfred saw them, too, and we excitedly got ready to land.

I had learned a few things since my moose hunt with Paul Carlson. Instead of landing in full view of my quarry, I ducked around the corner of the lake.

"Alfred, I'll taxi slowly past the point. As soon as I stop, you jump out and shoot."

Everything was set-up and everything went as planned. The duo came into view and Alfred bounded out with his rifle poised. What happened next was not expected. Unlike the Thanksgiving moose that ran, the caribou pranced toward us like race horses!

"Don't shoot yet! Let them come closer," I cautioned Alfred.

I couldn't seem to untangle my seatbelt fast enough. Once out, I crouched beside my partner. The pair approached within 75 yards. Alfred couldn't wait any longer. He squeezed off a shot and one caribou dropped dead in front of us. Alfred froze, speechless. We were both perplexed by this no-stalk kill. As luck would have it, the remaining caribou was not scared off by the shot. I instinctively raised my rifle, but then remembered the sun wanted to settle in for the night, and we couldn't dress out two caribou before the early afternoon darkness. Instead of taking the shot, I taxied the plane in closer, so we would have less distance to pack the meat. In spite of the engine racket, *Prancer* lingered curiously. As for "packing out the meat," it was within arm's distance of the airplane.

"This hunt was too easy," I remarked to Alfred. Which *was* true if we discounted the airplane drama.

When we landed, for the third time in the same day, it was mid-afternoon. The sun had held its head up just long enough for me to see the familiar airstrip. I flashed back to the second landing. Alfred must have been thinking the same thing.

"Ski okay." He reported.

Anna was at our house when I walked in. She and Ruby spoke at once, trying to tell me the story behind the scenes. Their words came out in interrupted jumbles.

"And then I prayed for a miracle," Anna animatedly explained.

"She'd heard what happened from CAA," elaborated Ruby.

"I just ran to see Ruby, didn't even knock on the door, and burst in the kitchen!" Anna laughed apologetically.

"She asked if I was still in shock." Ruby said. 'Shock about what?' I had no idea what was going on."

For a moment, Anna, Ruby, and I just looked at one another in amazement.

Ruby broke the silence, "I guess what I didn't know didn't hurt me."

Shooting the caribou had been simple, but nothing else about that day had been trouble-free. Ruby wrote my parents that it was the Miracle of 1958. Simply stated.

CHAPTER 11

KING OF THE ARCTIC

March 1958

IN A MATTER OF SECONDS, the J-3 lost 1,000 feet of altitude as a downdraft sucked it toward the coastline along the upper edge of Alaska. White-capped waves lashed out at our landmark shoreline that met the low 500- to 800-foot hills from where the tremendous unstable wind emanated. The tiny J-3 was no match for the severe winds, and in spite of full throttle, I could not hold the plane on altitude.

"We've got to get out of here!" I yelled to Leonard Lane, a large Inupiat (In-OO-pee-at) Eskimo, who was folded into my small backseat. We had just left Point Hope, the village where Leonard had grown up. I turned the airplane with the wind and let it drift away from the hills and toward the rough open water, where in contrast the air remained stable. The plane temporarily leveled off, but the next 20 minutes were a nightmare.

The howling wind increased in velocity and hurled gusts of snow down the barren hillsides and up into the air. At the same time, ice fog from the open ocean water gathered in patches around us. Wind whistled through the cracks in the plane's door and sent shivers down our backs as the temperature dropped from 20° to 0°. After 15 minutes of battering, we were down to 200 feet, searching for a point of reference and finding only sparse tufts of brown shoreline grass. New snow fell rapidly. The grass vanished. White out.

Panic! I figured we'd bury ourselves in a plane coffin. Before giving up the fight, I said a fast prayer and decided to backtrack to where I'd last seen the grass markers. I banked the plane sharply, all the time hoping I wouldn't become disoriented and fly the plane into the ground. As if a hand briefly lifted the white curtain, I could distinguish dark spots.

"See the grass, Doc?" shouted Leonard. "Level us out!"

We discovered ourselves over a lake crowded next to the shoreline hills. The raging blizzard allowed us no time to circle the "landing strip" or do a touch-and-go to test the landing conditions. I couldn't waste a second. I lowered the plane into the white blowing expanse, uncertain of the altitude, and having little depth perception. The 40 mph crosswinds threw the plane off balance. It hit hard on one ski, then skidded into and bounced off the hidden drifts near the lake's edge. Even though we were down, my fighting instinct remained strong. We were still not secure. As we dug into the two-foot snow for tie-downs, the wind shoved the plane around. Without losing sight of the plane, Leonard struggled toward the barren lake edge to explore the possibilities. Giving up, he improvised an anchor by packing a bundle of twigs deep into the snow.

At 2:30 p.m., somewhere north of Kivalina (Kiv-a-LEE-nah), we squeezed back into the plane to wait out the storm. The claustrophobic white faded into darkness and we packed our sleeping bags around us. All night the 60 mph winds hurled around us, and the plane trembled like a leaf.

We shifted our positions and tried to stretch our cramped bodies, but sleep didn't come naturally. There was plenty of time to think — and even talk, if we raised our voices.

"Hey, Doc. Looks like your dream turned into a nightmare!"

"We've sure had our troubles." I heaved a sigh.

Ever since I'd come to Alaska I'd dreamed of the King of the Arctic — a polar bear. Leonard, who worked at the Tanana hospital, had volunteered to accompany me on this expedition. He had the traditional and innate skills of Alaska Natives. Life-defying temperatures, subsistence lifestyle, and tundra terrain were as well-known to him as tornados, tumbleweeds, and farming were to me. He had all the makings of a perfect guide. I couldn't believe my good fortune.

For three weeks prior to the actual trip, we'd checked aviation charts and flight gear, experimented with engine heating equipment, and discussed how to spot a polar bear.

"I know it sounds impossible to find a white bear on white ice," Leonard had said, "But they are usually spotted by their shadow and their yellow color."

His informal teaching sessions included both mini-lectures and hands-on practice. To the delight of Naomi and Ruth, and the fascination of Ruby, we built an "igloo" out of the crusted snow along the riverbank. This would be critical if we were stranded without shelter. Leonard showed me how to cut the large snow blocks and balance them inward and upward to create a dome. Unlike the myth that Eskimos lived in these snow houses, the igloos were used for temporary outdoor survival, much like a tent.

Leonard Lane building an "igloo," February 1958

"Doc, we've done all we can," announced Leonard one evening. "Now it's time for the real thing."

On March 8, we soared off into the crystal-clear dawn. A 20 to 30 mph tail wind pushed us gaily down the well-marked course of the Yukon to our first stop, Galena. So far, so good. We refueled then set our course over untried country of rolling hills.

Finding our next checkpoint, Selawik (SELL-a-wik), was like playing hide-and-go-seek among the flat lookalike lakes in the vast whiteness. Even the Selawik River hid from us. After some time, and dropping to a lower altitude, we picked out the village and landed on the river in front of it. As in all the villages, a plane's arrival was a highlight. Villagers greeted us, timidly but curiously, wanting to know the *who, what, where* of ourselves.

After stretching our legs, we compacted ourselves back into the plane and flew a short distance to Noorvik. Once again, we landed, this time on the Kobuk River in front of the village. A small mob of about 50 rosy-cheeked Eskimo children surrounded us with bashful laughter, and then followed us to a cabin where one of Leonard's relatives lived. We stayed an hour, drinking thick, hot coffee and visiting. I'd never been much of a coffee drinker, but the substance was warm, and on the plain wood table there was even a tin cup with sugar in it to mellow the brew.

Another skip and a jump and we were at Kotzebue. I remembered my first experiences there with the tundra taxi. I thought of baby Andrew and pictured him as a chubby-faced child running around his village in a fur-ruffed parka and tiny mukluks. Even though that had happened only three years earlier, it seemed millennia ago in my experience and knowledge of Alaska.

We didn't take time to visit but refueled and pressed northward along the bleak arctic coastline. This plane sure didn't hold much gas.

White barrenness stretched in all directions. Trees shrunk out of sight to our right and massive ice packs appeared on our left. Before long, we passed over the sod houses of Kivalina. Herds of caribou spread out across the low hills. A myriad of frozen freshwater lakes followed the coastline, separated from the salty ocean by only several hundred yards of tundra.

"What's that down there by the ice crack?" asked Leonard, nudging my shoulder and pointing out the window.

With hopes high, we spiraled down, only to disturb a seal, content with sunning itself until the sound of the engine's roar sent him slipping back into the water. Polar bear didn't come quite that easily.

Six hours after we'd left Tanana, we sighted our destination — Point Hope. This village, balanced on the tip of a long gravel spit which extended into the ocean, was reported to be the oldest continuously occupied area in North America; with indigenous people living there for over 2,500 years. The peninsula, which offered good access to marine mammals with easy boat launching, was an attraction to whalers in the 1900s. Given its precarious location, the village's history had known tidal waves, storms, and flooding.

We bumped down on the snow-drifted tundra, and Leonard's family and friends swarmed around us.

"Doc, this is my brother, Amos," said Leonard. "He's the best hunter around and can give us all the latest hunting information."

"Yes, you've got a good chance of getting a bear," smiled Amos. He was shorter than Leonard, but had the same broad face. Just like Leonard, he didn't seem to mind that this White man wanted to use his hunting grounds. "Several bear were taken just last week."

Everyone seemed eager to talk to us. They were a congenial group.

"You can stay with us," volunteered Fred Fisher, introducing himself and his wife, Joan. They were young school teachers. He had dark, short clipped hair and enthusiasm written on his face. She was a tiny woman with features that resembled those of the Native villagers. I appreciated their welcome mat that came with accommodations and assistance.

After a while, we broke away from the small crowd and Leonard took me to the general store. We walked past a mix of tundra sod houses and frame buildings. The tundra houses, covered with snow, truly looked like stereotypical igloos. After wandering around in the store and only seeing bare minimums of flour, tea, canned goods, matches, army issue wool gloves, a deck of playing cards, and one bottle of Breck shampoo, it was clear to me that we did need some assistance.

"Leonard, what are we going to do about plane fuel?" I said, caught off-guard by the lack of fuel availability. All I could find was Blazo, a white gas, and kerosene. I had carried some aviation gas with me, but I hadn't anticipated this problem.

"Hmm," he said, staring down at the snow-tracked wood floor. The wheels in his mind were obviously at work. "Do you think the plane could fly if we mixed Blazo in with the remaining aviation fuel?" he said contemplatively.

I shook my head doubtfully. I didn't like this solution, but I couldn't come up with anything better. In preparation for the next day's hunt we hauled the Blazo to the plane.

Later that evening, Dan Lisbourne, president of the village, came over to the schoolhouse.

"Dr. Gaede, you might be able to use the 86-octane combat gas that was left here when the army company nearby closed several years ago."

I paused for a moment. That sounded workable, but it was too late now to do anything, so we waited for the morning.

March 9 wakened us with a bright brisk minus 8°. We hastened to drain the Blazo mixture from the plane and replace it with combat gas. As we emptied the tank, we found a leak in the gasoline sediment bowl. By the time that was fixed, it was already 1 p.m. At last, the hurdles were out of the way. Leonard packed his large frame into the backseat of the plane and I hand-propped life into it.

As impatient as I was to get on with the hunt, before heading over the open water to the rough ice pack, I circled the village, testing the gas in the engine. The plane swallowed up the gas, ran smoothly, and with full power, climbed into the motionless arctic air. All systems were a go.

We proceeded west at about a thousand feet, along the numerous small open leads of water. Every few minutes we spotted seal, shiny dark against the white ice. They never stuck around for a close inspection but slid into the water at the sound of the plane.

Leonard coached me about the strong old ice and thin new ice.

"The old ice often appears rougher and has a gray color, but new ice is smooth and whiter. Thin ice may appear slightly blue," he continued. "It's tempting to land on a large, smooth, pure white ice pad near an open lead to go after a seal or bear, but you must never forget the consequences of a plane breaking through the thin ice."

With that caution in mind, we scrutinized the area below.

"Do you think we're near the International Date Line?" I asked, trying to visualize the broken map line dividing the Russian and American territories.

"I think so, Doc. Can't you see the *red* (Russian) ice?" Leonard chuckled.

After no luck and a gas indicator slowing its bobbing, we returned to Point Hope. We discussed possibilities, and after refueling, played a hunch and began scouting southward along the coastline. About 12 miles west of Cape Thompson, Leonard hollered. "Doc, that's old ice ahead. We should find bear."

As we neared the first smooth field of old ice, I let down to 500 feet.

Suddenly, Leonard grabbed my shoulders, "There he is! Over there!"

I banked the plane and took a good look. Sure enough, just ahead was a light, yellow spot with a dark shadow. We made a wide circle. The bear sat back and, like a dog begging for food, raised his thick front feet with padded paws. Balanced in this way, his small head rotated as it followed the plane. Leonard assured me that the ice was old and could hold the plane. We landed on a flat area, downwind about a half mile from the arctic king.

"Doc, these bear think nothing of swimming for hours in the frigid waters; we don't want to spook him away," Leonard cautioned.

As fast as we could, we loaded our rifles and quietly crept along and over the six- and eight-foot broken pressure ridges. It was a confusing maze.

"Which way do you think we should go, Doc?"

At the same instance I caught a glimpse of movement to our right. We ducked and kept under cover as we stalked our potential prize.

Unexpectedly, the bear sensed danger and began running parallel to us. He was about 300 to 400 yards away — a tough, if not impossible shot.

"I'll get my sights on him," I whispered loudly. "You yell, and get his attention."

I poked my head up and I rested my 300 H&H over an ice block, trying to get the bear in my five-power scope. Leonard also tried to focus on the bear. Then he yelled. Hesitantly, the bear stopped to look at us with his black expressionless eyes. The front half of his body was exposed past an ice block, so I squeezed off my shot.

"Good shot, Doc! You hit him!"

We both leaped over the pressure ridges running toward our game — all the while Leonard attempted to focus my 8 mm camera on the bear.

Unexpectedly, the bear staggered out from behind ice blocks about 100 yards from where we had last seen him. Bright blood covered his front shoulder. Leonard stood back to photograph this arctic adventure. As I walked ahead, the white fury confronted me. Even though he was severely wounded, he defied me with open mouth and outstretched claws. I expected my next shot to end his misery, but the bullet placed in his heart area left him still fighting. After the third shot, everything became silent, except for the slow clicking of the movie camera.

Elmer's polar bear at Point Hope, March 1958

The beautiful heavy silvery-white bear appeared to be about four years old and was over eight feet from nose to tail. I stood admiring the king, but then realized long gray evening shadows were creeping along the pressure ridges, warning us of our limited time on the ice. Hastily we took a few more pictures then went to work. Leonard skinned while I flew the plane closer to the bear. Leonard was nearly finished when I returned with the plane. I removed a can of gas and packed the 60-to-75-pound bear skin in the back of the plane.

The can of gas and skin were not an equal exchange. We were overloaded. I studied our course. Uneven ice crisscrossed by large snow drifts paralleled several two-foot-wide cracks in the ice. Climbing into the plane, we hoped for the best. I opened the throttle and the plane picked up speed. At 40 mph, we hit a snow drift that bounced us several feet into the air before we slammed back down. Instantaneously, we struck another drift. We bounced 20 to 25 feet into the air. Still, we didn't have enough airspeed to remain airborne, and the plane whined in protest.

I figured I had another chance, but then nearly froze at the controls when I realized we had used up most of our runway. Tall pressure ridges sprung up in front of us. Just below us, I saw three fresh ice cracks with open water. I kicked a hard right rudder before we hit the ice. The plane swung diagonally over the dangerous cracks and then we bounced hard on one ski, and over to a flat area. With the plane still running, I opened the plane door and looked out at the skis.

"It's a tough plane," I admitted to Leonard, pulling the door shut, then discovering that the door window frame was bent.

Our second attempt at takeoff was successful and we winged our way back to Point Hope in the golden dusk. That evening as we related our hunting story, Leonard's brothers and their wives fleshed out and cleaned the bear skin. I couldn't believe how rapidly the women worked with their round-knife ulus. I also couldn't believe I'd survived the day with combat fuel, the arctic king, and a wild ice-pack takeoff. I'd wanted adventure, but today's experiences would last me for a while. I hoped the rest of the trip would be less harrowing.

March 10 dawned clear and calm, with indications of another beautiful arctic day. We couldn't have asked for better weather on this trip, and I knew it was unusual to have so many good flying days. All the same, I expected the weather to hold, so instead of immediately returning for the bear meat on the ice pack, I stayed in the village to participate in a yearly custom. In preparation for the whaling season, the Eskimos held a whaling feast.

At mid-forenoon, about 15 villagers gathered at one edge of the village over a large underground cache. My movie camera clicked away as a six-foot flipper of one of last year's whales was pulled up. The 300-pound chunk of meat was taken by sled to one of the houses and exposed for thawing in preparation of the next day's feast. I was amused to see a team of nine Eskimo men pull the sled while a woman in a bright red parka *mushed* them from behind the sled runners and was followed by a single dog.

At noon, Leonard and I backtracked to the ice pack to retrieve the meat — and the gas can.

"Doc, it's got to be here some place."

Searching the labyrinth of ice packs, we passed right over the bear remains without seeing them. Finally, I started a wide circle, tightening down until we saw the bloody ice. Knowing how wild an ice pack landing could be, I was relieved to make a less rough than expected landing.

The sun shone brilliantly and even though the temperature was 0°, the reflection from the pure white snow caused us to toss aside our parkas and shoot more pictures among the deep blue columns of pressure ridges. Like two boys at play, we jumped over the ice cracks and looked for seal holes, until Leonard warned, "Hear the ice creak? This stuff could move any time. Let's get over to where the plane is."

We removed the emergency gear in the plane and piled in the bear meat. Leonard stayed on the ice as I returned to the village and unloaded. While I shuttled back for Leonard, his brother's families prepared a feast of bear tongue and neck meat for us. Leonard and I knew we needed to leave the hunting scene and get back for the feast, as well as to stop testing the limits of our safety on the ice pack. Reluctantly, we packed the plane and left the awesome, but treacherous and unforgiving ice.

We'd planned to wash the bear skin in one of the beach water holes that evening, but when we walked to Leonard's brother's house, we found that his father, Jacob Lane, had already finished the washing and the skin was stretched across tall whale bones to dry.

Dismal overcast and a 15 to 20 mph wind opened our day of March 11. Just as Leonard had predicted, the ice pack had moved out during the night. Even with the foul weather around us, instead of trying to get ahead of the storm, we stayed in the village for the whale feast. The whale was cut into small portions and divided among the people. Leonard offered me some muktuk, the outer layer of the whale, but I had difficulty tolerating the odor, let alone getting it anywhere near my mouth.

During the feast, I had the rare and unexpected privilege of seeing our adopted daughter, Mishal's, natural grandmother, Beatrice Tooyak (TOY-yuck). Unlike many adoptive parents, who don't know much about their child, we'd learned that Mishal's mother, Dora, had come from Point Hope, where many of her relatives remained. Even so, as a physician, I was bound by confidentiality about adoptions, even my own, and I didn't tell Beatrice that her curly-haired, energy-plus granddaughter now lived with us. I did, however, capture her on my movie film so someday I could introduce Mishal to her.

Near noon, we bade farewell to the kind villagers, thanking them for their hospitality and fuel. We left most of the meat with Leonard's family, and since the bear skin was now clean and dry, we managed to fit everything into the plane without being overloaded. We gassed up and packed our plane, expecting to arrive home within hours. Our course had followed the coastline and now we climbed to 2,000 feet to cross over Cape Thompson. That's when the trouble began and the downdrafts threatened to destroy our fragile craft.

Now we sat in the darkness, wondering what would happen next and hoping there would be a sequel to this adventure story.

After a fitful night of intermittent sleep, we awoke to find that the wind had shifted to the opposite direction, and a gentle ocean breeze currently brushed over the enormous drifts, which partially covered the plane's tail and immobilized the plane. We urged our constricted bodies out of their paralyzed positions and nearly fell out of the plane into the deep snow. We stretched our arms above our heads, and then swung them back and forth to limber up our bodies then assessed the damage and our potential for take off. The end four feet of the fuselage was filled and packed with snow. The only way to clean out this dangerous dead weight was to stick our bare hands through the inspection plate holes. When our hour's work was completed, our hands were frozen and we pulled out the little Coleman stove to warm them.

Elmer trying to warm the plane engine oil with the little Coleman stove, March 1958

By noon, the ceiling had lifted sufficiently for takeoff. But even though the weather was cooperating, before we could attempt to start the plane, we had to tug it out of the white mire and pack down a takeoff strip. Finally, we managed to get off our drifted runway and fly to Kivalina, which was only ten miles away. The weather threatened us again, and heavy snow forced us to stay at this village. As before, the schoolteachers, this time Mr. and Mrs. Bingham, extended their hospitality. As isolated as they were, they were magnets for news, as mundane as it might be to someone else, and wanted to know where we'd come from, the success of our hunt, who we'd talked to, the conditions of the ice pack, and any other information we'd accumulated our expedition. After spending the previous night cramped into the plane, we rated their modest accommodations as five-star.

By the morning of March 13, the snow had subsided and we anticipated getting home that day. Our confidence disappeared once we were in the air. Much to our dismay, we butted 30 to 40 mph headwinds toward Kotzebue. It took us two hours and fifteen minutes to cover the mere 85 miles. Between the weather and limited-capacity gas tank, we weren't making much progress.

We landed at Kotzebue and, without wasting any time, refueled, checked the weather, and struck out toward Galena. Eventually, the wind decreased and we flew over the white flat tundra toward our next checkpoint, Selawik (SELL-a-wick). In like fashion as the start of our trip, we had to search for the scattered grove of trees that identified this isolated village.

"Leonard, I think we can still make it in tonight," I said optimistically, as we flew southeast with a benign crosswind.

I'd spoken too soon. At the first ridge of low mountains, incredible headwinds of 80 mph attacked us with the dual opposition of dangerous downdrafts.

"Hang on!" I shouted, trying to keep the plane under control.

The J-3, with its puny 75 hp engine was no competition for the gale-force winds. After being hammered for nearly half an hour and seeing the same terrain below me, except from varying altitudes, I decided to fly northeast in a crosswind to alleviate the stress of the wind beating on the plane, and wearing me out mentally. For a while, this new tack seemed to be satisfactory.

"Doc, do you know where we are?" inquired Leonard, as darkness caught us off guard.

"I'm not sure. I haven't seen any checkpoints for quite some time."

"We'll need to land sometime soon. What's your plan?"

Using the Common Sense approach, I followed a tributary to a fairly large river. It was 6:45 p.m. The gas gauge had ceased its merry movement, and the river was outlined in darkness. This "landing strip" looked long and wide, but I was wary about my actual distance from the surface and about obscured pressure ridges. Unfortunately, there were no good choices. I started down tentatively. To my surprise, and enormous relief, and after all the trouble we'd run into, I got a break. The air blended with the snow, and I never even knew when we first touched down.

We both let out our breath in unison and then climbed out of the plane. We fell into hip-deep snow.

"I guess we'll need our snowshoes." Leonard stated the obvious.

Before darkness completely shut us in, we tramped out a runway and found firewood. The roaring fire warmed us one side at a time and sizzled the processed pork Spam in our skillet. I was happy for something warm — and familiar—in my stomach. By this time in our marathon, I didn't think I'd ever tasted something so good.

"I thought you said we'd get *in* tonight." Leonard playfully fisted my arm and laughed. "I didn't know you meant *into* the plane."

At least we didn't have wind battering us. We climbed into our cocoons and prepared for another night out — or *in*, as it was in our case.

It wasn't hard to get up early after spending a night squashed into the plane. We flew down the river and 20 miles later we unexpectedly came upon Koyukuk, where we were able to refuel. It seemed the J-3 was always running on an empty gas tank.

Once we knew where we were and the weather decided to stop thrashing us, we were able to find our way to Galena and then back to Tanana.

In addition to numerous stories, some of which would be later verified by my movie pictures, I bestowed upon my family bear meat and souvenirs. Ruby, who would try anything once, reluctantly prepared large bear steaks. The odor forced us to suggest Leonard give the rest of the meat to an Eskimo family in the village. I could understand why Ruby preferred the Eskimo sunglasses made from shiny caribou hooves, a whale bone ceremonial mask, and a black-and-white ivory watchband.

A week later, Leonard walked into my hospital office with a newspaper in his hand. "Hey, Doc, you made the news." He pointed to a short article, which read:

Thursday, March 20, 1958

64 POLAR BEARS KILLED, ARCTIC — 34 FOR TROPHIES

A total of 64 polar bear were taken since January in the Arctic, according to Stanley Frederickson, U.S. Fish and Wildlife Agent, who with Sig Olson, Wildlife Manager, returned this week from a tour of the Arctic Coast area. Of the 64, Eskimos took 30 and trophy hunters 34. Of the latter, three were taken by dog teams out of Point Hope, but most by plane off the three-mile limit.

CHAPTER 12

BREAKUP TAKEOFF

April 1958

LIGHTS. ACTION. I wiped off my camera lens.

People congregated in clumps along the steep banks of the Yukon, nudging one another, laughing, and even placing bets on the anticipated drama. Children threw rocks and sticks down the bank, aiming for crumbled pop cans. Even though ducks and geese were returning north to announce springtime, and on this particular spring day the temperature had soared to 50°, something more was drawing young and old out of hibernation.

One might assume it was the annual river breakup, for the riverbank lineup was the same, but that was yet to come.

For Cheechakos, sourdoughs, and Natives alike, the thundering river breakup was something no one ever quite got used to. No, sir, each year at the Nenana Ice Classic, started in 1917, there were bets placed on the day and time the Tanana River would go out in Nenana, above Tanana, and start the sequence of breakup down to the Yukon River. In Tanana, breakup could happen anytime from the end of April to the end of May. The year before, the Tanana River had jolted loose in Nenana on May 5 and shoved through Tanana on May 16.

Whenever the river breakup took place, the entire village was notified. Sometimes, an individual would run to the Episcopal Church and ring the bell. The previous year, it surprised us at 3:30 a.m. and I asked the nurse on duty to blow the fire whistle. When the announcement came at night, people scurried out of their houses in bathrobes, nightgowns, and stocking feet for a "come-as-you-are" riverbank party. If it occurred in the daytime, school was dismissed for a real-life science exhibit. At the hospital, nurses, and even some patients, ran out the door to join the celebration.

Ruby described the massive event in a letter to her parents:

Yesterday, the ice started to move in one large mass at 3:30 a.m. Can you imagine a mass a half-mile wide and a half-mile long moving at once? It is a thrilling wonder! After a half hour, the ice jammed and then this morning at 6 a.m., it started again, and the large mass moved forward. For the next couple days we can expect mammoth ice cubes, all cramming together and floating together, covering the entire surface of the river.

In anticipation of the big event, the Natives watched the river for clues, such as the rotten ice in the middle of the river and freed anchor ice along the edges of the river. When these conditions were evident, the village men would return from trapping wolverines, red fox, rabbits, beavers, and wolves; both so they'd be on the village side of the river, and also in readiness to transition into summer activities.

In the fall, anchor ice was the first to freeze, whereas in spring it was the first to break loose. The anchor ice extended downward into the riverbed, forming a solid mass approximately four to six-feet thick. Reaching toward the center of the river, it would catch floating river ice chunks, and anchor them until the entire ice flow would freeze in place.

During the winter, the water flowing beneath the ice decreased, shrinking into the riverbed 10 to 20 feet away from the shore and the anchor ice. Then in spring, the melting snow would increase the flow, causing the river not only to fill its original bed, but to overflow and cover the anchor ice, until, buoyed up by the pressure, it would snap loose. The broken anchor ice would then float downstream a short distance before jamming together with unbroken anchor ice.

Currently, there were sections along the river with muddy channels of water10 to 20 feet wide between the shore and the main ice. Just the week before, the anchor ice had popped loose, and now the villagers expected the imminent arrival of the new fishing season and were repairing fishing wheels, fishing traps, and drying racks.

Not only did breakup signal fishing time, it also meant time to catch winter firewood. Reluctantly dragged away by the mighty Yukon, the ice clawed at the banks, desperately seeking a firm grasp; instead, it found and tore away tottering, root-exposed trees. After the ice had majestically thrashed and plowed down the river, it lost itself in the slate-gray Bering Sea — some 600 miles away.

In its wake, the toppled trees and sandbar driftwood gaily paraded down the river. The Natives joined the parade, racing around in their kicker boats, dodging the debris, snagging firewood, and then towing their catch back to shore to dry out. Cutting and stacking came later.

Breakup also heralded the reopening of the rivers as transportation passageways.

But breakup had not happened. Even so, the community had dropped everything, and people collected across from the hospital. All eyes were riveted to the river's edge. What was the attraction? Sandwiched on the shore, between the steep bank and the open channel of water, crouched a

Super Cub — on winter skis. The Cub was a heavy-duty version of my J-3, with a higher useful load, increased fuel capacity, and higher cruise speed. The waterway stretched approximately 40 feet wide and several hundred feet long. That morning, word had traveled quickly that the pilot planned to takeoff from the 50-foot strip of crushed ice and honey-combed snow. The same questions kept rolling around in the crowd: Will the plane crack-up on the shore? Will it end up in the water? The odds were formidable.

Why, I wondered, was the plane still there? A month earlier, I'd moved my plane away from the anchor ice, off the river, and onto the runway outside the village. But, somehow, like a gone astray snow goose missing its signal to migrate south, this plane had missed its cue to leave while the strong, wide river ice provided a long, sturdy airstrip.

It wasn't as though the plane had not received advance warning of its disappearing security. Two weeks earlier, Wally Hansen, my lab technician, who had come to Alaska from the Eastern Seaboard, had curiously set off across the river to the island, a half-mile away. Wally was not adept at reading the tell-tale sign of impending breakup: rotten ice. For example, he did not realize that the sun melted the surface ice, while the swift current, with similar intent, eroded the ice from beneath. To a newcomer, the resulting one-foot crystallized "candle ice" looked trustworthy, when in fact it could not consistently sustain a man's weight.

Unaware of the danger, Wally had confidently trudged through the worn-out snow on the river and around the winter-aged pressure heaves. The sun had beamed brightly as he had leaned against a slanted ice wall to rest at midpoint. Out of the blue, his peaceful world exploded, and his feet dropped beneath him! Frantically, he had struggled in the paralyzing water. With no moments to spare, he flailed his arms above the ice, grasped a firm chunk of ice, and dragged himself out of the frigid blackness that had wrapped itself around his legs, threatening to tow him away to the Bering Sea.

A wiser Wally beat a fast retreat back to shore — as fast as an adrenaline-powered man can cautiously crawl. His experience was a lesson for me: nature's clues needed to be learned and heeded.

"Here they come," yelled one of the bystanders.

The children stopped throwing rocks, and there was a momentary lull in conversation. Out of the hospital walked a man confidently swinging a suitcase followed by a woman carrying a baby. Down the gravel incline they marched toward the plane.

Who would be so foolhardy as to attempt flying off the token airstrip — much less, endangering two other people?

I knew the man and the plane. It was none other than Don Stickman. Don, a bush pilot, was a living legend — at least to this point he was living. I wasn't sure of his past, but I'd heard that he had served in the Air Force and had thousands of flying hours to his credit. Don lived downriver, but his escapades traveled upriver through my patients or filtered into the hospital whenever Don happened to be in the village. The middle-aged Athabascan Indian had a reputation of being able to fly in any weather, at any time, under any circumstances: ice fog, blizzard, total darkness, 30° below zero. But, could he get his plane with skis out of this predicament?

He did generate terrific Alaska Bush stories. For instance, in the middle of an examination, a patient asked me, "Say, Doc, you know what Don is up to now? Wolf hunting."

Bounty hunting could provide a substantial means of income for some Natives — without endangering the species. Most of the world's wolf population resided in Alaska, Canada, and the Soviet Union, with an estimated 5,200 to 6,500 in Alaska. Bounty hunting and trapping only approached 15 percent of the wolf population.

My patient went on to describe Don's hunting strategy. He flew low over the treetops, spotted the wolf, and then flushed it out into the open. His backseat gunner leaned out the open door as Don rolled the plane over for a clear shot. After landing and retrieving their bounty, Don would search for another wolf, while at the same time checking his controls, watching for a place to land, and hopefully staying above the treetops.

That winter Don scored 125 wolves, which sold for $50 bounty each, plus the sale of their pelts. Good money. Tricky flying. And tales visible on his airplane.

One afternoon Wally inquired, "Doc, have you seen Don's plane?"

Between patients, we slipped out the hospital back door, rather than walking through the waiting room and alerting patients or other staff that we were playing hooky. It seemed appropriate that Don's plane convalesced below the riverbank in front of the hospital. Tape and baling wire bandaged the engine and injured tail.

Now Don was preparing for another adventure; and I was concerned about whether or not he'd have any more after this. Ten days back, Doris had delivered a healthy baby girl. Yesterday she'd told me, "Dr. Gaede, I want to go home now."

"How will you get back to your village?" I quizzed her.

"I think Don Stickman is in the village," she said, as if the matter was settled.

"Yes, he has been here about two weeks now," I agreed. "But forget his help. There is no way he can takeoff from where he's parked."

"Dr. Gaede, I know he's on the beach," she smiled reassuringly. Her black braids bobbed up and down as she nodded her head emphatically. "Really! Dr. Gaede. I'll be okay with Don."

I felt responsible for Doris. My concern for her safety outweighed the thrill of capturing this event on film, so I handed my movie camera to Wally, who had fallen in step beside me. Then I scrambled down the bank to Don.

"Don, how are you going to take off?" I peered into his eyes intently.

"Don't worry, Doc, I've done this before." He grinned.

"Don, the gravel will tear up your skis, plus you have two other people to be responsible for." He didn't seem to understand the gravity of the situation.

"Hey, Doc, just watch," he said, tugging his cap brim down.

Seeing the determination on his face, I realized that there was as much chance of changing his mind as that of a spawning salmon swimming upstream.

A cloud darkened the sky as I surveyed the situation. The plane's skis were settled into the disappearing snow. Doris cooed to her just awakened baby. Don casually leaned against one of the plane's struts. The characters of this drama seemed oblivious to the nearby swirling water and the crushed river ice. I reluctantly returned to my filming position and consoled myself that most of the hospital personnel were literally on standby in case of an emergency.

Questions multiplied as the people's tone changed from excitement to pitches of concern. *"Doc, what did he say?" "Do you think he can do it?" "Boy, I'm glad I'm not his passenger!"*

Through my camera lens I narrowed in on the plane. Don, with the help of two other Native men, pulled the Super Cub to the edge of his runway to gain the greatest length possible for takeoff. Doris climbed in. He handed her the baby and suitcase. Then he slowly walked around the plane, touching a spot here and there, moving the flaps, and turning the prop. Apparently satisfied, he crawled in and fastened the door. Riverbank chatter stopped.

"Boy, that's a short strip," I mumbled to no one in particular.

"And the water's cold," said Wally, knowingly.

Don started the engine. While it warmed up briefly, I could see through the plane window as he checked the instrument readings and verified proper operation of the controls. He then advanced the throttle and brought the engine to full power. The plane roared and skittered across the snow. The tiny patch of snow passed quickly behind him and he was left with nothing ahead but water. He was deliberately heading for it! If the racket hadn't been so deafening, the gasps of the crowd would have been audible. Don was going to water ski! Could the 20 to 25 mph ground-speed hold him above the water? Was the water runway long enough? Would he get enough lift before smashing into the ridges of ice a short distance down the river?

Powerfully, the plane skied onto the water, leaving a wide wake. The skis supported the plane's weight until it lifted off gracefully. No one said a word. Eventually it became a dark silhouette against the pale Alaskan sky. Incredible! As the sun shone warmly and the ice continued to melt, the crowd, which had stood paralyzed, finally moved about. Looking at one another in incredulity, they clapped their hands. Slowly the spectators dispersed to fishing repairs, the general store, the hospital, and the CAA compound.

I stood alone and looked over the empty gray and white set. The only remaining signs of the edge-of-your-seat thriller were double ski tracks slashed through the disintegrating ice patch and a widening wake of gray waves gently lapping against the gravel shore.

Once again, I was a believer. Don could fly anywhere, in anything, at any time. And, if anyone was to question me about the truth of Don's stories, I had film to show them the facts. What a plot! What an action line! What a set! Tanana might have lacked televisions and stage plays, but it didn't lack the stuff movies and dramas were made of.

Fairbanks Daily News-Miner,
Thursday, February 26, 1998
Obituaries
Donald J. Stickman Sr.

Donald Joseph Stickman Sr., 77, passed away peacefully at his home in Galena on Feb. 23, 1998. He was born in Nulato, Alaska, on Nov. 21, 1920. At the age of 24, he was the first Athabaskan commercial and instrument rated pilot in Alaska. Donald was a veteran of the Army Air Force and served as a radio operator with the 42nd Troop Carrier Squadron during World War II. Donald raced in the Seward Marathon in the 1950s, a race he ran four times and won twice. He graduated in 1946 at the top of his class from the Army corps in Dallas. Back home, he established his reputation as a daring pilot who would go to great lengths to help others…

CHAPTER 13

HUNTING: NOT FOR MEN ONLY

September 1958

IT WAS THAT TIME OF YEAR when the early morning frost beckoned the golden leaves from the birch trees and the geese headed south. Crisp air pulled away the sunshine's warmth and a wool jacket felt more comfortable than scratchy. A touch of dampness in the autumn air brought out the woodsy fragrance. It was moose hunting season. On this particular year, the moose appeared plentifully along the Yukon River, and we could hunt the same day we were airborne.

"Ruby, hunting is not for men only," I cajoled my wife one evening when I was cleaning my gun. "You're a rugged Alaskan woman and a strong-stomached farm girl."

I thought this was a fair statement. It was true in the sense that she was unafraid of guns, comfortable with killing animals for food, liked moose meat, and enjoyed the great Alaska outdoors.

"I really don't have time, and I surely don't want to stay out overnight," she countered.

"We'll just go down the river a few miles, spot a moose, land on a sandbar, you shoot your moose, and we'll return by dark." I was building my case.

"I don't know." She twisted her hands in her apron and stood her ground.

Tanana was in the middle of terrific moose country so this really was a simple matter. Essentially, it could be as uncomplicated as driving to a grocery store.

The previous fall I'd flown out, shot a moose, dressed it, and been back to see patients — all in the same morning. No big deal.

Rutting season had just begun and I'd been told that a bull moose could be called easily during this time. I'd walked down to the Arctic Missions house-chapel to discuss the subject with Roy Gronning, a more experienced hunter than me. The bulky Scandinavian said it actually was true — and that he'd demonstrate the next day.

The 6 a.m. air was exhilaratingly cool and made for good hunting. The J-3 skimmed off the river, and we winged our way west under a low ceiling along the low-slung mountain range between the Yukon and Tanana rivers. Flying at 500 feet we saw several grown bulls within the first minutes of flight. Lakes were plentiful and I chose a long, narrow water strip. The J-3 settled effortlessly on the mirror that reflected the yellow leaves of surrounding black-and-white skinned birch trees. I taxied toward the shore and a flock of ducks flapped and squawked out of our path.

Roy and I both wore hip boots and sloshed around getting our gear ready to set-up. Roy stayed near the plane and I stationed myself 50 yards along the shoreline behind a stand of birch. He'd stashed a large piece of moose rack behind his seat and now put it to use, scraping it against a small spruce tree. With a little imagination, the noise was that of a moose walking through a heavy growth of trees. I wasn't sure why that reverberation "called" a moose, but I suspected that any bull within hearing range assumed his territory, or harem, was being threatened and he needed to defend his interests.

Bulls will deliberately rub their antlers on small trees and brush to make noise when they think they are responding to another bull's intimidation. In no time they can de-limb and peel a six to seven-foot spruce tree in the process. Sure enough. Nearly immediately following Roy's call, I heard an identical sound in the distance, perhaps a quarter mile away. Roy continued a rhythm of antler rubbing and waiting. The bull out there was definitely interested and we could hear him coming our way.

After hearing some thrashing about, I caught a glimpse of him. Two hundred yards away, in scattered scrub and trees, the brown animal, with an exceptionally large beard, kept moving toward us. We were downwind and well hidden; all the same, he stopped about 75 yards from me. To my advantage, he'd sensed danger in Roy's direction and stood at a diagonal to me, exposing his shoulder and neck. I slowly lifted my 300 H&H and rested it against a tree branch. Looking through the scope I put the crosshairs just behind his front shoulder, hoping to minimize damaged meat. I squeezed the trigger. The monarch buckled and lay still. Roy and I motioned to each other and came out into the open. I put another bullet through the bull's head.

He measured seven feet tall and nine feet long, nothing exceptional as far trophies go, but I wasn't out for a trophy. I was after meal-making. Thirty minutes after we'd taken off, we had a real meal deal. We gutted, skinned, and quartered the freezer-filler. By 9:45 a.m. I was back at the hospital.

The two-seater J-3 had been at gross weight with Roy, gas, hunting gear and myself, but I'd packed in a front quarter of meat anyway. After work, I flew back for the rest of what would be roasts, steaks, and moose burger — totaling 650 pounds. The meat would be preserved in the hospital locker and shared with the Gronnings.

This fall, I wanted to take Ruby moose hunting. I'd started on her again and thought maybe helping her dry the supper dishes would enhance my persuasion.

"I just don't know," she balked. "Who will watch the children?"

"One of the nurses or Anna Bortel."

She scrubbed extra hard on a plate and bubbles flew in the air.

"I just don't know," she said again. "The weather has been overcast... and you know how I feel about landing on the beach, and..."

Not many dishes were left and I was short on time.

"Now Ruby, you know I no longer have a puddle-jumper. My new plane is nearly first-class."

Parting with the J-3 hadn't been easy. When Mark, two-and-a-half, had watched the Christmas-tree-colored J-3 take off from the river without me, he'd cried, "That's my plane, it's a nice plane, too." He'd been my constant flying buddy and felt a kinship with the J-3, perhaps like a child with a favorite pet pony. One winter day Ruby couldn't find him. He wasn't tussling with a moose rack or trying to ride his trike in the snow. She walked across from the house, and looked up and down the banks of the river. There she spotted the short-legged guy in his red parka, against the white snow, and beside the J-3. She marched over to the aspiring pilot.

"What are you doing here?"

"I'm checking Daddy's plane." A beatific smile spread over his chubby face and his blue eyes shone.

Ruby's eyes didn't shine; nor was she smiling. Along the way, she'd picked up a stick and now prodded him up the bank and back indoors.

Of course, I was amused, and a bit proud, when I heard the story. My son shared my love of aviation and also seemed to have a lot of confidence in me. Another time when Ruby was telling Margie Gronning, the missionary wife, about a harrowing trip back from a flying house call, Mark had chimed in, "My Daddy's a good pilot."

He had all the makings of a good pilot himself.

All the same, our growing family could not fit into the J-3, so that summer I'd sold it for a four-place PA-14 Family Cruiser. Both planes were Pipers. In comparison to the J-3's 75 hp engine, the PA-14 had 115 hp. Even though the factory specifications were 115, bush pilots frequently modified their planes to have larger engines. The purpose was not speed, but quicker acceleration for short strips. A limited number of PA-14s were manufactured, so they were a rare find.

I'd flown the new plane, on wheels, from Fairbanks. My family knew I'd be coming in. When they heard me buzz the village, and saw the crimson underbelly, Ruby loaded up the red wagon with Mark and Mishal and started to the airstrip. Naomi and Ruth ran ahead. In a rare show of excitement, Ruth jumped up and down, ran around the plane, patted its red finish, wanted to climb inside, and begged ardently for a ride. I figured out later that behind all this fanfare was her satisfaction that we could fly together as a family; which we did that same evening. The girls and Mark tightly crowded together on a flat bench seat. They strapped the single lap belt across all three. In front with me, Ruby held Mishal. This was Mishal's first airplane ride, and I didn't know what to expect from the restless five-month-old. Initially, her head bounced about in curiosity, but then she settled down and acted as though she'd flown all her life.

I was elated, too, with the luxuries: an inside starter so no longer would I have to hand-prop my plane; a radio to hear weather reports en route or call for help; and landing lights.

The new red Family Cruiser with Grandma Agnes Gaede, Ruth and Elmer. Ruby is in the back seat, 1958

In spite of all these upgrades, I shared Mark's sad feelings. As I'd stood on the banks of the Yukon, watching that small plane with large memories become a speck in the broad Alaskan sky, I felt especially nostalgic, remembering my experiences with it on floats.

The J-3 had been a great fisherman's friend. Take that Saturday in August, when I'd conferred with Roy's about fishing.

"Roy, where's a good spot for catching sheefish?"

"Let's try the Tozi River," he suggested. "It's only 11 miles down the river."

We flew down the Yukon and, in a few minutes, landed at the mouth of the Tozi. I carefully edged the floats over to a coarse gravel beach and docked. We stepped off the floats in our hip boots, into the shallow water, and toward the hollowed-out bank, where we cast off with our daredevil lures. It wasn't long before we had not only a string of sheefish, but northern pike as well.

"This ought to take care of a few meals," said Roy, as he hopped onto the float and swung his long legs into the plane.

I pushed the plane off the sandbar and jumped onto the right float as we drifted out into the six to eight mph river current. While standing on the float, I hand-propped the plane. Immediately, the engine fired and I climbed into my seat. Run up seemed normal, so I pushed the throttle open and prepared to take off. That's when I noticed something was wrong.

The plane seemed glued to the water. Acceleration was very sluggish, and the plane did not come up on the "step." Curiously, I looked out the right window. A-okay. I checked out the left window. Where was my left float? I couldn't believe my eyes. There it hung a half-inch below the water. Making a preliminary diagnosis of a hole in the float, I aborted takeoff and returned to the shore.

We dragged the plane as far out of the water as was possible and started a more thorough investigation.

"Here's the problem," I explained. "The middle compartment in the float is nearly filled with water. We must have 30 to 40 gallons in there."

In a half-hour, I'd pumped out the water and located the offender, a quarter-sized ragged hole in the thin aluminum near the center.

"Do you have any ideas for improvising a patch?" I asked as I rummaged through my meager emergency gear.

"I'll show you a bush technique for boat or float emergency repair," Roy answered, then turned and walked off along the shoreline.

A few minutes later he returned with a cut off half-inch alder branch. I watched him without asking questions. First, he took his blue handkerchief, reached inside the float compartment, and packed it into the offending hole. Then he cut the branch and placed it against the hanky. Pressure held it in place as it wedged tightly against the opposite side of the float. Finally, he closed the float cover.

"This should get us home."

Once again, I taxied out into the river. Both floats rode normally and I expected to be home within minutes.

I was wrong. As soon as I took off, the throttle went wild. The engine uncontrollably wound up to full throttle. I managed to circle the river, switch off the engine, and plunk down on the water. With my emergency paddle I maneuvered the plane to shore, and a second time we made an emergency inspection.

This time, the throttle cable had broken, leaving me without power control. The farm boy that I was, I figured baling wire could fix anything, so of course I carried it in my emergency gear. I jury-rigged the fuel control with two wires: one to open the throttle and the other to close it. Then I threaded the wire from the carburetor into the cabin through the outside of the cowling and door edge.

"Roy, ready for Take Three?"

I taxied back out into the river. This time we made it home.

That was only one of the many experiences with the J-3. I hoped I'd make good memories — with positive endings — with the Family Cruiser as well.

Returning to my task at hand, I continued to plead my case. "Ruby, I even have tundra tires that are large, soft, and designed to accommodate planes taking off and landing on rough terrain."

"Okay. Okay," she said in exasperation, pulling the sink plug and letting the water gurgle noisily down the drain. "I'll try anything once."

And so, the chain of events began just as I had calculated. One evening, Anna stayed with the children as Ruby and I flew 15 miles down the Yukon.

"Look, Ruby, there he is!" I pointed to a moose bedded down in a dried swamp on an island less than 200 yards from the river. He turned his head and sunlight flashed off his large 65- to 70-inch rack.

By this time, I had wheels on the plane, rather than floats. We landed on a wet, slightly soft sandbar about 600 feet long. I knew the moose waited for us between the swamp and our sand bar. We'd find him in deep grass and autumn-red bushes just beyond a thick grove of spruce trees. This would be effortless. We smeared ourselves with mosquito dope, and I put on my army-issued Yukon packboard. Its design of plywood slats with hooks along the outer edges for lashing on a canvas bag or wild game was heavy, yet functional, and with no comfort whatsoever. Along with this cumbersome necessity, I carried my 300 H & H. Softly we crushed through the incredibly dense underbrush, over large, flat, moldy-looking mushrooms and around velvety, moss-covered stumps. In the shadows of the trees, light snow dusted low-lying flaming crimson cranberry bushes. Tightly gnarled alders held us back and five-foot-tall grass obscured our view. I struggled through the thicket with Ruby tucked close behind me.

"What about bear?" Ruby whispered apprehensively. She looked over her shoulder. "You've told me they're in tall grass like this."

"Don't worry," I said, as we broke onto a narrow trail, which I hoped wasn't a bear trail. I kept my ears trained for the sound of their warning "woof."

After about 15 minutes, I cautioned her to be silent. Any moment I expected to come out into the yellow swamp and see her giant bull moose. Slowly the woods parted and we looked around.

"This... looks familiar..." Ruby paused.

And then I saw it. I couldn't believe my eyes.

"Well... I'll be..."

There in front of us was the airplane. We'd walked in a complete circle, pulling toward the right, which was a very common error in overcast or evening skies.

"I made a slight miscalculation," I stammered. "Let's try it again."

"Not tonight," she said. "I am *not* going back in there again. I want to go home."

The light rain was on her side — and her square jaw was set — so we walked out of the wet scrub over to the plane. As we walked, I observed the shore and how our feet squished into the fine sand. Even with tundra tires, the drag would increase the necessary distance for takeoff. Furthermore, the engine was underpowered for this kind of strip.

I tried to appear nonchalant. "I need to do a bit of preparation for takeoff," I said. "Go ahead and get into the plane."

Then, with my hiking boots sinking, I walked to the end of the short strip and found a six-inch thick, long length of driftwood. This would be perfect. I placed the log directly in front of the water and secured it into the wet ground. If I couldn't pull off by the time I reached this point, I'd hit this springboard and the plane would bounce up and gain enough loft to continue into the air. At least that's what I'd learned from the sage advice of "hangar talk."

After getting this in place, I walked back to the plane with what I hoped was a semblance of calm. I climbed in beside Ruby, and in what I hoped was a "by the way" tone, I said to her, "Here, hold the flashlight so I can see the speed indicator. When we hit 45 let me know and I'll pull off." I hated to include her in this fly-by-the-seat-of-your-pants bush takeoff. I knew it would do nothing to build her confidence for a return hunt.

The plane rolled along the sticky sand and sluggishly gathered speed.

"Forty-five!" she called out — without *any* semblance of calm.

I used first notch of flaps and pulled back on the stick. The plane lifted off, but hovered over the water barely above stall speed. At treetop level, I retracted the flaps and congratulated myself for not needing the extra boost from the log. It had been nip and tuck, but we'd made it. Regardless, I would have to get a bigger engine before trying a stunt like that again — or I would lose all my passengers from fear. Ruby never said a word the entire way back. As could be predicted, immediately upon landing in Tanana, she very tight-lipped climbed out of the plane and walked home in silence, alone, while I tied down the plane. Yes, indeed, I'd traumatized my wife.

I don't know what made me do it, but several weeks later, I actually heard myself saying, "Ruby, what about another try at moose hunting?"

I'd hoped time would have blurred the nerve-wracking experience of our first attempt. I blundered along, "I've been scouting around and there's a bull with his harem just a couple of miles upriver. In fact, it's so close to home that we can be there in a few minutes."

She didn't say a word. She just sat with her mending on her lap, threading a needle, and secured the knot with an extra-strong tug. Perhaps enough time hadn't elapsed to dim her memory. Deciding that this was a lost cause, I turned to leave.

Then she stopped me with her slow words, "Elmer, you know how I feel about your flying ... and I don't know much about hunting. But if you *promise* not to take any chances again, I'll go."

In high spirits I made the rounds at the hospital and contacted Olga Neufeld, one of the nurses, to see if she could stay with the children. At noon, I threw the hunting gear into the plane, the 300 H & H for myself and the 30.06 for Ruby.

Ruby was dressed for the hunt. No matter what she wore or how hard she worked, she always appeared feminine — even in *my* hunting clothes which were designed for someone five-foot ten-inches. She'd turned up the brown canvas coat sleeves and stuffed the denim jeans inside her rubber boots. A stylish green wool-blend neck scarf, or "muffler" as we called such neck wraps, kept the mosquitoes off her neck and accented her hazel eyes. A trace of lipstick remained on her lips.

We took off toward the confluence of the Yukon and Tanana rivers. Just as I thought, the moose I had previously spotted were still feeding in a narrow meadow off the side of the Tanana River. The bull lay surrounded by his three fair ladies.

"How does he look to you?" I pointed out his distinguished masculine broad rack.

"He looks like meat for our table," she remarked, without much passion.

"Okay let's go after him." I tried to buoy her attitude with my cheerfulness.

I searched for a landing spot and found a 60-foot wide empty slough about a quarter mile from the moose. Although the dark sand indicated wetness, I decided that my tundra tires could land without a problem. I didn't want to spook the moose, so I pulled full flaps and glided quietly onto the sand. The plane rolled smoothly for about 150 feet, then stopped abruptly. Ruby climbed out of the plane and immediately sank several inches into the sand — and it wasn't as though she was heavy by any means. I climbed out beside her and we both stared as water filled the tire tracks.

"We'll never get out of here," she said in alarm, and her face looked as if a cloud had just covered the sun.

I believed that if I could walk through the mud that I could fly the plane off it; but rather than argue the point, I took her hand and helped her climb up the bank to firmer ground. Things were not starting well.

We walked to the small stream of water separating us from the moose, and much to our dismay, found it to be too deep to cross.

"Don't worry, we will find your moose someplace else," I assured her.

Together, we pushed the plane out of the gray muck, and I taxied up to more solid ground where we took off and flew downriver 35 miles. There, along a beautiful, firm shore, stood a moose.

"This could be it, Ruby," I said, dipping a wing to see if it had a rack. Sure enough. Short spikes. I landed evenly, and we both jumped out with our guns. The moose started to run. In the excitement I forgot that it was

Ruby's moose. I pulled off a shot, but didn't know if I had hit it. He splashed into the stream and then crashed off into the woods. I didn't like to leave a possibly wounded animal, so we hiked after him. The thick undergrowth and tall trees thwarted our good intentions and I called off the hunt. Even when we took off and flew low over the area, we could not find the bull.

It was now 3:30 p.m. and I resigned myself to defeat. Hunting with Ruby had become more and more of a disaster. On the way back to Tanana, I decided to check if the other bull had remained in the meadow with his pleasant companions. Just as I suspected, he was there. Nothing would be lost with a final attempt and even if she didn't want to shoot the moose, I could. I thought more carefully about my landing strip and found a drier area — which was on the same side as the moose.

Trying to redeem myself in her eyes, as a real hunter, I checked my compass and scrutinized every landmark. As we walked through the woods, we climbed over fallen timber, staggered around mini bogs, and pushed through tall grass. I didn't know how the wildlife managed to keep track of themselves. The towering spruce trees refused to let the sun in, and on several occasions, we stumbled into faded brown swamps with spindly swamp spruce, and were surprised by the sudden brilliance. Ruby bravely followed along, apparently suppressing her fears of bears and being lost. I had to give her credit. She wasn't complaining, and we were in some rough country and both carrying heavy guns. I considered myself to be in shape and even I was getting a workout.

I suspected the bull was around the bend of heavy brush, about one hundred yards ahead. We edged forward, hugging the brush along a large corn-husk colored meadow. I could smell him. Standing up and leaning forward, I broke cover. There he was, looking right at us. Without delay, he tossed his antlers and lowered his huge head. He was going to charge! The ground shook as he pounded toward us. I backed up and nearly knocked Ruby off her feet.

"Get ready!"

The moose picked up speed. Ruby froze.

"Shoot, Ruby! Shoot!" I yelled.

She stood paralyzed in his path. By now he was only 50 yards away. Too close for comfort. Frantically, I focused my gun on the monster. Just as I pulled the trigger, I heard another shot ring out. Only 37 yards away from us, the moose crashed to the earth. I didn't know what was trembling more, the ground from the impact, or Ruby as she turned to me with terror on her face.

We both stood gasping for breath.

"You did great," I managed to choke out. "Now finish him off."

She brought her rifle back up to her shoulder and with two shots stilled the quivering animal. My heart pounded and I could nearly hear Ruby's. I imagine she, like I, replayed the previous minutes in her mind. She'd had every reason to be scared stiff.

Ruby and her moose, September 1958

Already this season, I'd gutted three other moose, so I immediately went to work gutting this 900-pound hunk of meat. Ruby had never seen this stage of moose-hunting, although she had cut up and packaged pounds of meat after they had been hauled home. When she recovered her sense of speech, she observed with curiosity the innards of the moose.

"He's like a camel," she said in amazement. "Just look at all that blood and liquid. And look at his heart — the size of my head."

I knew she was comparing him to the cows and pigs she'd seen butchered.

The evening darkness and gnawing mosquitoes hurried us. I decided we couldn't complete our task.

"We can let him cool down overnight, and then tomorrow morning Roy and I will skin him and pack out the meat."

I hated to leave her trophy so abruptly, but as she'd said, she didn't want to spend the night in the wilds. To be sure that Roy and I would find her success story, we strung toilet paper on our way back to the plane, like Hansel and Gretel leaving a cookie trail.

Within five minutes of takeoff, we were back in Tanana. I was jubilant and raring to re-talk the hunt, however, Ruby wearily walked home. In silence. She put the children to bed and crawled into a hot bath. I expected she needed some time alone. If I ever wanted her to hunt with me again, I knew I'd better grant her that opportunity.

Early the next morning, Roy and I flew back. The weather was still sunny and the toilet paper hadn't disintegrated with moisture. Seven hours later, all four quarters of Ruby's moose were back to Tanana. This part of the hunt was familiar to her. She and I would be busy for many a night picking hair off the meat, cutting it into various cuts and sizes, and wrapping it for the freezer.

I was mighty proud of my wife's hunting adventure. Since I hadn't taken my movie camera along to document her story, I decided we should mount the head and rack of her moose. Subsequently, Leonard Lane and I made a return trip. Once again, I relied on his skills, and he obliged my request to skin out the head. Of all things, on our hike back in, we literally ran into two more bulls! Leonard shot one and the other went crashing through the woods to safety. Now we embarked on dressing out his — before we continued with retrieving the rest of Ruby's. By day's end, I figured I'd had my fill of moose hunting — for a while. Within just a few weeks, I'd helped four people get a moose. We had plenty to share with the school teachers and later for the Christmas village potlatch.

This was Ruby's first, but not last, moose hunt. She had proven she could bring home the moose and cook it, too. She never really took to hunting with the airplane, but later when we relocated and we could drive on back roads, she was willing, and even somewhat eager, to get up early, or drive at dusk, with two guns between us.

Ruby's Moose Travels

The head mount was first sent to Ruby's parents, Solomon and Bertha Leppke in Peabody, Kansas. Later, it was transferred to Elmer's parents in Reedley, California. After their deaths, the moose found a home with Elmer's brother, Harold, in Fresno, California. In 2015, it was transported in Dan Doerksen's fruit truck from Reedley, California to Soldotna, Alaska, where it is back with the original Gaede family, on the Gaede-Eighty homestead.

CHAPTER 14

CLOSE ENCOUNTERS OF MANY KINDS

September 1958

"DOC! What's your stall speed?" shouted Pete.

I yanked my focus away from searching for white critters and stared at my control panel. My airspeed had dropped to 44 mph — almost stall speed — and I was about to nest in the treetops. Instantly, I applied full throttle, pushed the stick forward, and barely cleared the trees.

Pete Miller, the hospital maintenance supervisor, had joined me that morning when I'd pointed the Family Cruiser north toward the Ray Mountains. The thick-built man with a dark crew-cut was eager for new scenery, other than in the village, and accepted my invitation without hesitation.

From the air, clumps of yellow and yellow-orange birch dotted the variegated green spruce, letting us know the all-too-short Alaska summer was coming to an end. We flew over Allakaket (alla-KAK-it), positioned on the Koyukuk (KOY-yuh-kuck) River, up the Alatna River, and into the Brooks Range. Low-lying purple-red foliage now replaced the golden birch, and the pale gray granite gave the illusion of snow on the mountainsides. In other areas, the real snow reached down toward timberline. Swirls of clouds decorated the sky above the picture-perfect landscape below.

Following the river into this seldom-hunted area, we spotted dozens of moose, their antlers reflecting the sun as they ate breakfast in the meadows. Occasionally a black bear splashed in the river to check out its own breakfast possibilities. Before long we spotted the *white* we thought we were looking for. A fly-by along a sheer mountain ledge turned the specks into a herd of about 20 sheep.

Neither of us had hunted sheep before. "I imagine there must be a legal ram somewhere in that herd," I said. We couldn't rely on each other to evaluate the potential.

Suffering from amnesia about my other hunts, I had no question that we would land, hike over to the sheep, find the ram, and return home successful hunters — all in a day's work.

The Alatna had no suitable sandbars on the mountain side of the river, so we looked for an alternate landing area. One potential landing strip appeared to be an old dry riverbed. To get a closer view, I'd used first-notch flaps, slowed the plane to 55 mph, and skimmed the trees. Not realizing the deceiving rise in the elevation of the riverbed, I'd slowly pulled back the stick to near stall speed. I'd been too busy to notice my predicament until Pete barked his panic about the stall speed.

After several more passes, I was confident the riverbed, with scattered two-foot aspen and spruce, would make an adequate 800-foot landing strip. We could dodge the small trees, and my durable tundra tires would take care of the rest. I passed over a trophy-racked moose, lowered my flaps, and started down.

To my horror, the *small* trees were growing up to 10 and even 12 feet! Too late I realized my error. I wanted to put this scenario in rewind and start over, but there was nothing I could do now. I was committed to land.

The early airplanes didn't have full harnesses for protection, only a lap belt. Many a pilot had planted his face in the metal front panel.

"Brace...." I yelled.

Then we hit. Like the threshing machines I'd grown up with in Kansas, the plane frantically threshed the leaves from the trees. For what seemed like an eternity, we were thrown around violently. Then it was over. The leading edge of the left wing encountered a ten-foot tree which pinned us in place. We turned to each other slack-jawed. Amazingly, we and the plane were still intact. For a brief moment, I felt like laughing.

The PA-14 in the bush landing, September 1958

Pete and I climbed out to survey the environment that had ensnared us.

"I think we just initiated your new plane," observed Pete, eyeing the gold-green aspen leaves adorning the red struts.

"And I just put on a new tail wheel last night."

He ducked out from under the wing and turned around. His eyes glanced upward. "Doc! Look at this."

I ran my hand across the front of the wing. "Do you think I can tell Ruby I hit a bird?"

We walked along the perceived *smooth* riverbed, which in reality was gravel-packed, bumpy, and sloping. Unlike the trees that had stretched in size, the 800-foot airstrip had shrunk. At the most, it was 300 feet of "axe and shovel work," after which it fell off into a 12-foot gully. Anyone searching for us would have a tough time spotting us in this unwanted camouflage.

Alaska is known for its bush plane graveyards scattered over the vast tundra and mountain passes. Sometimes the pilot makes it out, sometimes planes are rescued and brought back to life; but more often than not, both pilot and plane stay where they crash until the wilderness gently, but surely, buries them from view. That's what we thought would eventually happen to the Gullwing Stinson.

I'd flown over the wreck the previous year, located downriver of Tanana, about a mile south of the Yukon River. The rugged, roomy, and robust workhorse was lying on its back in a patch of black-green spruce, its yellow fuselage standing out like a caution sign — warning of bad weather flying consequences.

Naturally, I was interested in the story of this powerful bush plane. As is the case in all tales, there are differing perspectives, facts, and hearsays. This is how I pieced it together. About six years earlier, Joe Cook, a bush pilot, had been flying alone. Foul weather had pushed him down for an emergency landing. *That* he managed. The next morning, Joe hacked at spruce trees until he cleared a short strip. Even with the timber out of the way, the underbrush clawed at the heavy plane. After several aborted attempts to take off, and more axe work, the crash-landed plane lumbered back into the air. There was hope — at least for a short while.

With Tanana nearly in sight, the engine sputtered. Flat gas gauge. So close, yet so far away. The plane went down hard and fast. This time the impact and tangles of scrub flipped it on its back. Joe lived to walk away, but the wheels-up plane was worn-out.

As usual, I found Roy Gronning to go with me on this exploration. I easily pinpointed the site from the air, and set down on the sandbar along the river. Before starting into the heavily treed area, now thick with a half-foot of snow, I took a compass reading. I wanted to increase my odds of walking in a straight line and not circles. Regardless of my precautions, along the way we strewed toilet paper. After an hour, our treasure hunt produced no reward.

"Roy, how did we fail to find it?"

I wasn't about to give up. We popped back up in the air. I could see our trail. We'd only missed our prize by a couple hundred yards.

The weather was holding and our curiosity did as well. We flew back to Tanana to recruit an assistant. This time, I dropped off both men for ground reconnaissance. My job would be to direct them from the air by flying tight circles over the Stinson. This worked. I landed, followed their newly papered trail, and the three of us assessed our find.

Much of the exterior fabric was missing, but the engine and cabin were intact. We salvaged several seat cushions and a small tank.

"We could pull out the engine," I said.

We weighed this idea. Did we really want to lug it through the woods down to the river? What value would it have for us? An interesting thought, but not practical. At least we'd investigated the possibilities and made an informed decision.

We left the Gullwing Stinson, expecting it would stay in its graveyard forever more.

We didn't want to stay here forever. For a while Pete and I worked to clear the airstrip. He was a short, sturdily built man, and chopped and dug steadily beside me. But, the autumn air and damp woodsy fragrance along with the cheerful sunshine spelled *hunting*. Time was wasting.

"Pete, let's worry about this later. We came here to hunt. Let's get going."

We started our ascent up the mountain. Luck was with us and we stumbled upon a game trail. This made hiking much easier than scrambling through unruly vegetation. The hunt literally seemed to be back on track. That is, until we rounded a corner and encountered another obstacle.

A full-grown black bear, plump in preparation for hibernation, lazily waddled straight for us. No indication of stopping. Our guns were strapped to our packs, and we really didn't want to shoot a bear. We banged our pack boards and shouted. I'd heard that black bear have poor eyesight. Obviously, this one was unusually hard of hearing, as well. He swayed toward us on a collision course. We clamored about trying to untangle our guns. Finally, at one hundred feet, he recognized his mistake, snorted, slowly veered off the path, and ambled around us.

"That was too close!" I burst out, expelling all the air in my lungs I'd been holding.

We continued winding up the mountainside, carefully picking our way across a broad section of slippery shale and onto a grassy mountain bench.

"I know we spotted the sheep right here," I said, puzzled.

We looked around. No sheep. We hiked another quarter-mile.

"Maybe the sheep have spooked into the crags," offered Pete. "Let's look over the edge."

We crawled to the edge of the bench and peered down. Only more ledges of gray rock. No sheep.

"They have to be here someplace. Let's just sit quietly and wait," I recommended.

They were certainly at an advantage since we did not have the freedom to climb where they could. Their gravity-defying safety was our danger.

Thirty minutes crept by slowly. We sat in absolute silence. White clouds grazed on the purple, windswept mountains to our north. Then as I was just leaning forward to discuss a new strategy with Pete, we were rewarded by a clatter below, followed by a single-file parade of 30 pure-white animals — most with one-eighth to one-quarter curls. The young ones turned their gentle black eyes and serious faces toward us, before nudging their mothers and trying to hurry along the procession.

"Hey, Pete." I whispered. "Where's the ram? Do you see a three-quarter curl? ... Are you sure these are sheep?"

We scrutinized the parade. The great white ram we sought was nowhere in sight. Baffled, we stood there discussing what to do next. A stiff gust of wind chased down our necks and helped us decide, as did the low hanging afternoon sun. We started down the mountain, back to the job of clearing an airstrip.

We worked hard. The chill left our bodies, and the major deterrents left the takeoff course.

"Doc, are we really going to get out of here?" Pete leaned on his shovel.

One thing was certain. With the abbreviated airstrip, we would not be able to take off as loaded as we had come in. I came up with a plan and Pete complied. Not that there was much else he or I could do.

"Let's empty the plane of gear and the ten gallons of gas. I'll fly solo over to the sandbar over there." I leaned back and stretched my shoulders.

One problem remained, a very big problem: the gully. Once again, I'd have the chance to try out the log technique. We dug a shallow trench and snugged down a six-inch-diameter log just before the drop-off. Just in case I didn't get airborne before the gully, my tundra tires would hit the log and I'd bounce into the air.

Hopefully the "suitable strip" on the other side would not be a mirage, as had been the riverbed. I would go by air, and Pete's journey would be by water. My companion shook his head, but couldn't come up with a better idea.

To transport him and the gear the 50 feet across the river to the sandbar, we lashed together some of the trees we'd chopped down and made a raft. Pete would wade with the floating gear through the shallow water.

With that part of the plan in place, we pulled the plane back into the trees as far as possible. I climbed in and Pete held onto the tail. This would be tricky — but our only chance to get out. I started the engine and gave it full throttle, while at the same time I held down the brakes and used first-notch flaps. The aircraft lurched about as the engine roared, eager to take off, but restrained by the brakes *and* Pete's arm hold. I couldn't see behind me, but bet Pete was getting covered with dust and torn-up leaves. I was praying hard.

Once released from Pete's grip, the light-weight aircraft bounced crazily over the rough gravel and tree stumps. The gully rushed toward me. Just a few feet before hitting the "last chance" log, the Cruiser wobbled into the air. Finding myself airborne never felt so good. I gained a bit of altitude and then blinked my eyes to be sure the new landing strip was as faultless as I thought it was. I'd been mistaken before. This time, however, the sandbar *really* was wide, flat and 1,200 feet long. I circled and prepared for landing. All went well and I touched down without incident. I climbed out and waved to Pete. I'd gotten myself out of the dead-end spot, now it was his turn.

I got ready for Pete by building a driftwood campfire. At least my cold-legged partner could warm himself after his walk in the river.

Pete tossed a rope across the river to me and then fastened the other end to our loaded raft. I was sure he would have no trouble fording the river. Slowly he eased himself down the bank and into the water. Then suddenly he slid into the icy water and dropped out of sight! What had happened to Pete? How could this shallow creek swallow him up? Was this, too, an illusion? In times of exasperation, I often uttered, "Oh for Pete's sake." This was truly the appropriate comment for the situation.

The raft dipped precariously from side to side. I yanked the rope. Pete, who had once been a lifeguard, bobbed up, howling about the frigid water. Fortunately, he had the presence of mind to grab the raft to steady our precious cargo. Arm over arm, I pulled steadily on the rope. None too soon, the raft and my shivering partner docked at the sandbar. Pete let me deal with the raft as he stripped off his wet clothes and threw himself at the fire's hospitable warmth. Between the two of us, we managed to find enough dry clothes in our packs to redress Pete.

This hunt had taken a course all its own, and now we'd used up most of the day. Pete's teeth finally stopped chattering enough to discuss what to do next. Should we spend the night here? Should we fly back to Tanana?

A commotion across the river interrupted our conversation. There before us was a prize bull moose, checking us out with his nose-filled face. As we stood staring, he plunged into the river and swam toward us. Uneven waves of water spread before him.

"I think we're being challenged, Pete."

Our sandbar was devoid of trees or any hiding place. I hoped the moose wasn't like a domestic bull and attracted to red, or my plane would be his target.

Neither of us wanted a moose. Our freezers were full, and regardless, we were 200 miles from Tanana and didn't have room in the plane with all our gear. We yelled and pounded our cooking pans together. The bull, with a massive 70-inch rack, continued its approach. We reached for our guns. About 100 feet from us, like the bear, it changed his mind and decided to detour around the two obnoxious creatures on the sandbar. We didn't take our eyes off him until he crashed back into the woods.

At this point, we could almost read each other's mind. Enough was enough. Delusions of airstrips. A shuffling bear. No ram. Pick and shovel sweat. Immersion in a mountain river. Belligerent moose. And now, light rain drizzled from heavy evening clouds.

We reloaded the plane, climbed inside, and enjoyed every foot of the long sandbar — and then the cabin heat. I turned the plane toward Bettles Field, about 70 miles away. There we could find some safe, indoor camping.

Visibility deteriorated rapidly and I followed the twisting river 200 feet below. In 30 minutes, the Bettles Field beacon, like a friendly lighthouse, welcomed us.

Later, comfortably full of moose steak, Pete and I sat in front of a crackling fireplace in a cozy, log cabin with CAA friends, telling and reliving our wild and woolly adventures of the day. The encounters with the gully, bear, and moose were clear in my mind, but I still wondered, *Did we miss him? How could the Great White Ram have evaded us?*

The Story of the Gullwing Stinson

Thirty-one years later, Naomi Gaede-Penner was reading "We Alaskans," a section in the Fairbanks Daily News-Miner. The January 29, 1989 article by Mike McCann described how he had rescued a plane, restored it to glory, and had it flying again. It was the same Gullwing Stinson that Elmer and Roy had inspected.

CHAPTER 15

HOSPITAL WARD THE SIZE OF KANSAS

October 1958

I'D MADE THE MOST OF SEPTEMBER, grabbed every inch of daylight, filled our larder with moose, tried out duck hunting, and added other adventures. Now, in no uncertain terms, winter was here. The frozen river quieted the landscape outside our living room winter and snow settled the dust of Front Street. Mail planes had changed from wheels to skis and there were more bad-weather flying days than good. I'd had plenty of fun in September, but that was a bonus, not the job description of serving at the Tanana PHS Hospital. Accordingly, I needed to make medical field trips within the area of my responsibility — before perpetual darkness and polar temperatures shut out the possibility.

I couldn't separate physician from bush pilot. The terms were inextricably wrapped together. For that reason, I assumed I'd fly my own plane for these field trips, rather than try to match my schedule to the variable schedules of commercial and charter flights.

My first stop was upriver to Beaver. The PA-14 purred, and the weather cooperated, but those conditions didn't last. On the way to another village, the engine fumed and balked. I messed with the fuel-air mixture control and the carburetor heat. Everything sorted out and I relaxed my grip on the stick. After a no-problem landing and take off, the disturbing cycle started again. This time I was near Tanana. I limped in, set down the red plane, and walked away. With the merciless winter weather, I couldn't take any chances. It would have to stay at the airfield until I could get a mechanic to diagnosis its malady.

As odd as it felt, the rest of the jaunts would have to be in a different plane — with a pilot other than myself.

My circuit extended both down the Yukon River and up the Koyukuk River. I didn't know when I'd be back. It wasn't as though I could pick up the phone and check in at the hospital — or let Ruby know I was okay. Supporting me in the background were proficient bush nurses and a capable wife, but still I felt the weight on my shoulders of my patients in the hospital, patients in the villages, medical staff, and family. To choose one sector, was to choose against another. I located a charter and flew out.

I documented these trips to the Medical Officer in Charge, Dr. Ernest Rabeau, whom I had first met in Kotzebue when I'd taken baby Andrew to Noatak.

MOC PHS ANHS Anchorage, Alaska
Attention: E.S. Rabeau, M.D.
October 16, 1958

Field Trip Summaries (Traveled to the first three villages with my own plane.)

BEAVER, ALASKA:
Left Tanana Sept. 30, 1958, 1:36 p.m.

Arrived in Beaver at 4:36 p.m. High winds, average 45 mph aloft — light snow, and blowing sand entire distance with severe turbulence to Rampart.

Stayed with Episcopal missionary, Reverend Paul Glander at a cost of $3.00 per meal and $3.00 per night. His wife, a Registered Nurse, takes care of the medical problems of both Beaver and Stevens Village.

School teacher is Mr. Ragan. School physicals were given morning October 1, 1958. Visits were made to the homes in the afternoon. Teeth extractions were done in the evening at the school. There is very little work in the village and most everyone is on some kind of aid. A government project building new cabins provides work for 7 men of the village.

The young girls seem to mature unusually early, with average age for menstruation about 10 ½ or 11 years.

Left Beaver enroute to Stevens Village Oct. 2, 1958. 9:30 a.m.

SUMMARY OF STEVENS VILLAGE, ALASKA:
Arrived Stevens Village, Oct. 2, '58 10:15a.m. with headwinds — 1200 foot ceiling and about 8 inches of snow on the field.

A new Territorial school is being built. There are no school classes now. Mr. Elmer Long is the teacher. I saw all the school children (20), and preschool children (13), as well as almost all the others

(22), in the shelter well which was their church. Reverend Donald Nelson is the missionary. I finished general clinic at 6:30 P.M. We only had 6 extractions. People are on some kind of financial assistance in general. Considerable drinking problems. Children in good condition in general.

Left Stevens Village 11:12 a.m. Oct. 3, 1958.

SUMMARY OF RAMPART, ALASKA:
Arrived Rampart 11:50 a.m. Oct. 3, '58.

There are about 60 people in the village. Clinic was held in the one class room Territorial school. Thirteen school children, 6 preschool children and 7 adults were examined. Five people were examined at the dental clinic with the result that 10 teeth were extracted.

The villagers earn their living mainly by fishing, and a little trapping. The storekeeper, Mr. Weisner also provides some work such as with his salmon cannery. The near-by gold mines occasionally use several men. Economically the village is about average.

Mr. Weisner provided nights lodging. On October 4, 1958, after heating the plane engine in the early 10° temperature, I had to make three runs at the take-off to get off the snow. There is a fair air-strip of about 2500 feet. The strip had 8 to 10 inches of loose snow. There was a strong head wind and I arrived back in Tanana at 9:30 a.m. October 4, 1958.

SUMMARY, HUGHES, ALASKA:
Left Tanana, Alaska at 9:30 a.m. on Wien's Twin Beech aircraft.

Arrived 10:30 a.m. There was a low ceiling and we had considerable difficulty getting through the village. Hughes has a runway of about 5,000 feet.

There are approximately 65 people living in Hughes. In the afternoon I examined twenty school children, in the one class-room Territorial school taught by Mr. Raymond Wilson. The children (nearly all) had severe dental caries. Also, during the afternoon I examined most of the families with the help of Mrs. James, who has the roadhouse. In the evening we did extractions. The next forenoon we examined more people and did more extractions. There were 12 dental examinations and 16 extractions.

The men of the village are exceptionally industrious and are good workers at the nearby gold mines. There is but little drinking and welfare assistance among the people. The striking part about the school children is their seriousness. I was told that hardly ever could anyone get the children to smile or laugh.

Mrs. James takes care of the local medical problems. Mr. and Mrs. (Leslie and Patti) James are the Wien's agents, have the post-office, and run the roadhouse. They have a great influence over the people and I believe they are a definite asset to the community.

Twenty school children, 19 preschool children, and 16 adults were examined.

Left Hughes Oct, 9, 1958, 1:30 p.m.

SUMMARY, HUSLIA (HOOS-LEE-A), ALASKA.
Alaska: Arrived Huslia, 2:05 p.m. Oct. 9, 1958. Chartered flight on Oct. 9, '58 from Hughes, with pilot Sam White on a 180 Cessna at the rate of $45.00 per hour. Time of flying was 45 minutes but we pay for Charter both ways. (Mail run to Huslia is weekly.) There is a 2,500 foot run-way and we were met by a truckload of people. I immediately went to the Territorial school house where I met with the school teachers, Mr. & Mrs. Ley Kahl and Mr. Ronald Anderson. The school is a three-class room type similar to the new Tanana school, but is several years old. That afternoon I examined the first and second grades. In the evening Reverend Pat Keller and I made house visits to elderly people. On October 10, '58 forenoon and until 2:00 p.m. I examined the remainder of the school children which was a total 59. From 2:00 p.m. until 6:00 p.m. I examined families. We had a high-school boy, Barney Sackett, bring the families to the schoolhouse, while Mrs. Kahl assisted me in my records. We saw approximately 60 people. In the evening I extracted 13 teeth. The next morning I had 7 extractions and saw several adults with medical problems. I had very satisfactory quarters at Lucy Sackett's Roadhouse. The charge was $3.00 per meal and $3.00 per night.

As a whole this village of about 150 people is a fairly progressive group. Most of the men work at the gold mines at Hog River, except during the winter months. There is some trapping, especially Beaver trapping. Moose were plentiful this season and fishing by nets was normal.

Many of the buildings were moved back from the river bank this summer because of the river bank erosion. The people have privies and the dogs are staked along the outer edge of the village.

The people appear relatively healthy except for many dental caries and quite a few younger children have rough dry skin suggestive of Vitamin (A) deficiency. Very few children receive vitamins.

On the next field trip visit to this village, I believe it would be appropriate to recommend doing refractions and using the audiometer. I also recommended to the village leaders to write to our Dental Officer, offering to pay the trip for our Tanana Dental Officer and his assistant.

SUMMARY OF KOYUKUK, ALASKA

Since there are no scheduled flights between Huslia and Koyukuk, I had to charter Mr. Zimmermen for this distance. He has a 4 place Stinson and charges $35 per hour. We left Huslia about 11:15 a.m. October 11, 1958, and arrived at Koyukuk at 12:05 noon. There is a 1600 foot landing strip.

Koyukuk is a village of about 120 people. The men work intermittently at trapping and fishing. There are considerable drinking problems... both men and women. The village doesn't appear as progressive as Huslia.

There is a two classroom Territorial school taught by Mr. and Mrs. Rison. There are 33 school children. Since I arrived on Saturday I did not examine the school children separately, but I examined all the families in the village as they came to the school house. During the examination I filled in the school health records. In the evening the teachers entertained several of us guests with a film.

The forenoon of October 12, 1958 was spent writing summaries and other paper work. I spent all day and evening with extractions and seeing general medical patients on October 13.

October 14, 1958, weathered in with rain and low ceiling. I did some paper work and saw several patients. In general the people were fairly healthy and several needed dental work and refractions.

Economically, I do not see any immediate optimistic future. There is a remote possibility that Global Oil or some other oil company may again provide jobs in their work in the Nulato (New-LAH-to) area.

Finally got off next noon on October 15, 1958 on Northern Consolidated. We stopped at Nulato a few minutes for mail and then went to Galena. I had a short layover before the plane arrived to go to Tanana. I went to the BIA school and discussed the village medical problems. I had only enough time to examine 3 people. The more serious medical problems are usually referred to the air force medic at Galena. I was not able to discuss the economic condition of the village but aside from a few jobs at the air force base I do not believe there is much work available.

I arrived back at Tanana about 3:00 p.m. on 10-15-58.

Although this village life and culture was familiar to bush school teachers, bush pilots, missionaries, and some healthcare workers, these field trips had taken me a layer deeper into Alaska and the lives of my patients. I came away with a greater insight into the backdrop of their lives, which was useful in more effectively serving them. It was frustrating that I couldn't solve the problems of health and economics. I could only do so much, which was to provide the best medical care possible. That was my desire.

The reoccurring problems with the PA-14 were finally remedied when I purchased a rebuilt 125 hp engine, which replaced the original 115 hp engine.

CHAPTER 16

TOOTH PULLIN' TIME

Winter 1959

"PLEASE DR. GAEDE — I'm brave." The somber Indian boy begged me with both his eyes and words. Nearby, a group of his friends flaunted the holes in their mouths, prior habitats of decayed teeth. "I'm tough," the thin, small teenager pleaded softly.

My anesthetic supplies were depleted and I knew his tooth could wait until my next visit to Kaltag, a Koyukon Athabascan village 200 miles down the Yukon River. All the same, I was keenly aware that this boy's request had surfaced not out of medical necessity but out of a recently assumed social need. As I toyed with the forceps, the silver metal took on a new shape. I was holding in my hands the key to his acceptance. How could he go home, the only boy that evening without the trophy wound to prove his manhood among his peers?

The boy's request had prompted laughter in the crowded schoolroom-turned-office where a baby cried, two preschoolers played tag with a husky pup, and several of the older men coughed chronically. Then there was silence as 20 to 30 villagers listened for my reply. The school teacher who assisted me, by holding a flashlight, did not assist me with this decision; she just looked at me without saying a word. I felt caught in the middle holding an uncomfortable power. Whatever I did, this young man would leave with pain — either of social rejection or physical wound.

Pausing a moment to reflect on my dilemma, I mentally paddled back to the fork in the river that had towed me into this deep water.

The last Public Health dentist had made his rounds two years before at Tanana and along the Yukon. Since that time, all those in need of dental care had to either fly to Fairbanks or remain in pain. Both arrangements were less than ideal.

I should not have been surprised when Mary Ann Burroughs, my spunky new director of nurses, urged me to consider this crucial field of village medicine, then took a step further and one day presented me with my first dental patient. She'd come from the Hollywood area, as a private nurse; when she wore her sunglasses and draped a silk scarf around her head, it was not hard to imagine how she'd fit in.

"Doctor, there is an elderly Native man in the waiting room. He has several large cavities and is in excruciating pain. Can't you help him?"

"Well, Mary Ann, I really would like to help, but I do not have any dental experience, much less dental equipment."

Reluctantly, she relayed the message to the old man and sent him home with medication.

I thought my logic would silence her plea. Instead, it triggered a search, and like a group of kids on a treasure hunt, she and others on the hospital staff spread out, searching the attic, basement, and old supply closets.

"Oh, Dr. Gaede, we found a whole drawer of dental instruments and something like a dental chair!" Mary Ann returned to my office with delight. It was as though she was preparing to stage a show. "And by the way, here are several books on dentistry."

I'd wanted to be a bush pilot and a bush doctor, not a bush dentist. Nevertheless, my patients' needs pushed me through the classroom doors to "Bush Dentist 101."

Fortunately, the medical terminology, anatomical descriptions, and techniques were easy to understand. My three years of giving major anesthetics made the dental blocks relatively simple. I presented myself with honors when I ended my studies.

A few days later, I made the announcement. "All right, Mary Ann, I'm ready. Prepare my first victim — I mean patient."

Putting on an air of confidence and experience, I walked up the stairs to the second floor of the hospital and into the recently designated "dentist office." It was a small office without windows and contained a small number of white cupboards. Sam, the Native man in distress, sat straight up in the black, cracked vinyl dentist chair. He had been living on aspirin and hope that I would treat him. Immediately he opened his mouth, pointing deep inside and groaned. My overhead light sought out the offending molar. Yes, I could identify and understand his discomfort. So far, so good. I asked Sam to close his mouth for a moment.

"Sam, I'm going to give you something so you won't feel me pulling your tooth."

Without a sound, he endured the pokes of the needle. As we both waited for the dental block to take effect, I asked him about his family and his trap-lines. Apparently, the winter had favored him with full traps.

Then, gripping the forceps, I slowly loosened his tooth. A strong yank sent the bloody culprit flying across the room. Mary Ann, Sam, and I smiled at one another in triumph.

"You'll feel a lot better now. Are any of the others giving you trouble?" I tried to be casual.

He shook his head in a vigorous "no."

My career as bush dentist was successfully launched.

Elmer as dentist, February 1959

As fate would have it, my fame spread up and down the Yukon River. One morning as I was making hospital rounds, I heard dogs barking madly at the front door. Before I could reach the door, a snow-covered Athabascan Indian stumbled inside, leaving his dog team yapping on the hospital porch. A thick wolverine ruff obscured his face. All I could make out as he mumbled through frozen lips was "Help...doctor...tooth."

Five minutes later, the thawed-out middle-aged man explained that a severe toothache had caused him to mush from Rampart, 80 miles upriver, to find a doctor. I removed the offending molar and then by his request, extracted two more teeth — as a preventive measure.

"Don't want to come back. Too cold and long," he told me. Pulling his ruff around his now-smiling face, he shook my hand, thanked me profusely, and disappeared out the door.

It's always good when a patient leaves smiling.

Naomi and Ruth smiled at my stories, but they were less enthusiastic about being the subjects of these accounts themselves. Since they were early grade-schoolers, it was time for their front teeth to make way for permanent teeth. Both were a squeamish about this process and wouldn't even wiggle their teeth loose. I'd take them up to the dental office at the hospital. Regardless of which was the actual patient, they'd come together. I'd cotton-swab on topical anesthetic and we'd talk a bit while it took effect. When it was time for the forceps, whoever was in the chair would squeeze her eyes close and hold her breath until the split-second ordeal was over. The tooth would then be carried delicately home in a piece of gauze and placed under a pillow at bedtime for the Tooth Fairy to bring a shiny dime.

I didn't think anything of this process until I removed Ruth's lower incisors. I expected that within weeks I'd see slivers of white gleam from her gums, but nothing appeared. There were no permanent teeth beneath the surface. This was a case where x-rays would have been useful and preventative.

"Doctor?" the boy's insistent question drew me back to the present. The sea of faces watched expectantly, and I felt the pressure of going ahead and making a decision. Wishing for the wisdom of Solomon, I reluctantly agreed to torture the boy at his request. Immediately the villagers edged around us, some with words of empathy, others shaking their heads with misgiving. No one wanted to miss out. The drama before them was most likely the best entertainment available in the village.

Adroitly the village schoolteacher focused her flashlight as I took hold of and slightly loosened the tooth. The room was silent. The boy winced.

"Remember, I don't have any medicine to keep this from hurting. You can still change your mind."

He shook his head, his eyes wide with anticipation. I loosened the tooth some more. Then, with a steady tug, the tooth departed from its socket.

The spectators cheered! The boy held his jaw. The other teenage boys rushed around him. He blinked hard and then a triumphant smile slowly crept across his face. Acceptance rights were his. I took the special tooth and placed it in a plastic container. Later I'd add it to my pint jar of other trophies.

The boy tugged at my sleeve. "Thanks, doctor."

I couldn't keep up with the dental demands in the village, which *were* extensive. My primary job was that of medical officer. Administration, deliveries, accidents, monitoring tuberculosis symptoms, and trying to provide medical assistance to the villages consumed my days, and sometimes nights. The school teachers, missionaries, and innkeepers in these villages did their best to fill the gaps through a "clinic" held via two-way-radio.

"Allakaket, this is Tanana radio! Allakaket, Tanana radio! Do you read?"

"Tanana radio, this is Allakaket. I read you loud and clear."

"Come in Allakaket."

"I've had people with terrible toothaches. Could you arrange a trip here for a day and teach me how to extract teeth?"

The voice was that of Dorothy, the nurse, who was also the wife of the Episcopal minister. I'd met her during another field trip. She'd emanated courage and dedication, and I observed the Natives' respect for her.

Allakaket was a small village 100 miles north of Tanana, on the Koyukuk River. During a lull in my schedule, and an opening in the weather, I flew up to train her. On the south side of the river, eight new cabins neatly lined the bank. On the north side, older cabins tumbled along the bank. Athabascan Indians lived on the south side and Kobuk Eskimos on the north, which was actually called Alatna.

Dorothy met me at the airstrip and we talked over general medical concerns as we walked to her clinic. When we opened the door, several elderly Athabascans sat up straight in anticipation. This dental lesson would be an immediate hands-on practicum. Class came to order. I reviewed the basics of dentistry and Dorothy watched intently as I extracted two posterior molars.

"Now it's your turn."

She was willing and confident. With a minimum of coaching, she did an aesthetic block and proceeded to select the correct forceps.

"Remember, you take a strong grip and loosen the tooth by rocking it, before you try to pull it."

Dorothy was an outdoorsy woman who probably did her share of chopping wood, wrestling moose quarters to be cut and packaged, and shoveling snow. Her strength showed as she applied the forceps to the tooth and tightened her grip.

Crunch! The tooth shattered.

Wildly she looked at me. Tears filled her eyes and she hugged the tired looking man whose mouth still gaped open. Fortunately, the anesthesia numbed his pain, but Dorothy's own ache from concern spilled over.

"What did I do wrong?" her voice broke.

"Don't worry. You just crushed the tooth. It must have had a weak shell. Maybe it is good this happened."

Her shocked expression didn't look like there was any good happening.

I explained, "Now you can find out how to take care of this kind of complication. You'll have to remove the root piece by piece."

Little beads of perspiration glistened on her forehead. She pushed back her bangs and took a deep breath. Carefully she cleaned the debris and swabbed the socket.

Her first patient got up, mumbled his appreciation through unfeeling lips, and seemed satisfied. His experience and the dentist-intern's trauma didn't seem to faze the two waiting patients and Dorothy logged additional dental hours.

After a while, and even though she never complained, Dorothy looked exhausted. I suggested she make herself a cup of hot tea and we'd sort through the dental instruments to decide which ones she could keep for her newly acquired dental practice. The warm beverage and much needed break restored her spirits.

"Thank you, Dr. Gaede, for teaching me how to help my people." She squeezed my arm.

"Her people," on both sides of the river, were fortunate to have this strong, tender-hearted woman. She'd do anything for them.

Sometimes I wondered what happened to the boy at Kaltag. Maybe his story was told, embellished, and retold until he became a legend in the village. On the other hand, perhaps he became a dentist himself. Whatever the outcome, he'd bravely withstood enormous pain at the expense of his pride. I thought back to that moment when I'd I looked him straight in the eye, shook his slight hand, and said, "No problem, son."

He'd gotten up and walked timidly among his admirers who pretended to slap his arm and then inspected his mouth.

I'd cleared my throat and started gathering my instruments. "No problem at all..." But the truth was, I really didn't like to inflict pain on any of my patients. I much preferred pain-free circumstances.

Huslia, Alaska, May 5, 1959

Dear Doctor Gaedy,

Is there eny tenetist over there right now. I sure would
Like to get too front tooth in. I get lots tooth missing
Every Sinch I was kide. . . .

Very truly yours

A Century of Faith: 1895 — 1995
Excerpt p. 115, Dorothy Mendelsohn

"The radio was most important to us because at 4:00 p.m. four days each week the doctor at the Tanana hospital (Dr. Elmer Gaede in 1957-1959) would make a round-robin call to all villages within his jurisdiction to inquire and offer advice about medical problems. His contact and concern were most reassuring when he came in loud and clear; when his voice became distorted or did not come through due to poor signal conditions, it was devastating. The same held true for attempts to make plane contacts in emergency situations. We were very dependent on that outside link and were often cut off from it. Those were the times our isolation was the hardest."

CHAPTER 17

DOUBLE FEATURE DRAMA

March 1959

FOR THREE DAYS ice fog had coated our world. Delicate frost crystals hovered on the thermometer, which dipped between 20° and 40° below zero, and the weather created severe electrical radio disturbances. I was concerned about medical situations in the downriver villages.

The static of the two-way radio had become an expected part of my routine. I attempted to ask the right questions, make a diagnosis, and prescribe over-the-phone treatment. In addition to keeping me in touch with the villages, the radio conversations connected the teachers and missionaries with a small piece of "civilization" — which illustrated how isolated the villages were in comparison to Tanana.

Winter in Alaska is harsh. Everything is locked in darkness, with perhaps a few dismal hours of midday grayness. Small village missionaries, especially women, live with extra tension. Their children's restless energy is penned into one or two room cabins — claustrophobic quarters when there is no escape outside. In the 1950s, schools were not available in every village and education often fell on the shoulders of these mothers. In most cases, the Calvert Curriculum was used, and, 70 years later, is still used by Alaska home-schoolers.

Elmer making radio contact with other villages in the Interior

Today, just as back then, if these women could venture out, there would be no place to go. No corner cafe for a cup of tea and stimulating adult conversation, no library for fresh reading material, no store with interesting browsing or shopping for birthday cards, sewing notions, spices, or a bright sweater.

In the 1950s, there were no televisions, computers, or telephones to provide contact with loved ones or to gather news of the world outside the village. Today, mail service is still dependent on weather, daylight, and flying conditions. Inevitably, these factors progress to "cabin fever," an expression for what happens to individuals who live without sunlight and social stimulation and stay inside for extended periods of time. Symptoms include depression, restlessness, and sometimes violence.

We were fortunate in Tanana that we could, however, get several radio stations with featured programs, along with news. Other information was carried into the village through pilots, or an occasional week-old newspaper. Political or world news drifted in through letters from friends and relatives, and the highlights were then passed around among hospital personnel, FAA families, and schoolteachers. (Civic Aeronautics Authority — CAA changed to Federal Aviation Administration — FAA in 1958.)

Ruby and I tried to support and offer encouragement to the missionaries. Our basement turned into a "Tanana Holiday Inn." Pregnant missionary women would come with their children to await the arrival of another child, or families would fly or boat in just to get out of their villages. Other times, we would fly down the Yukon on "missionary journeys" and carrying along news, some tangible items, listening ears, laughter, and fresh perspectives.

On this particular day, communication finally broke through with an inter-village telephone call to the hospital from the Tanana FAA station: medical crisis at Kaltag, approximately 200 miles downriver. The school teacher had desperately tried to reach the Tanana hospital by two-way radio; when that was not possible, they tried communicating through the Galena FAA, which was often capable of relaying information during weather difficulties. No luck. Still undaunted, Kaltag tried the Bethel FAA, which then leapfrogged the message to Tanana FAA, and then to our hospital. The message: *A child is in a coma and needs immediate medical attention. Can Dr. Gaede fly in?*

I sent back the reply: *Weather permitting, Dr. Gaede will be there as soon as possible.*

For two days I paced around the hospital waiting for the fog to lift.

Ruby wrote my parents:

Dear Loved Ones,

The past week was more normal.... I went to a baby shower.

Monday and Tuesday were busy for Elmer as he had a cancer patient which was dying. He had a premature baby (not an incubator here!) so he was not home much these days and nights, but he did what he could to prolong life.

Wednesday night he went to check at the hospital on an emergency at Kaltag (this was done by radio.) They said the patient was still in convulsions, so Elmer came to the house and said he'd get the plane ready for when the weather improved. I asked him why the patient couldn't be brought in, but he told me the planes went to Kaltag only once a week and then the connections were so terrible that a sick person would die before he ever made it here...

Love Ruby

P.S. The premature baby died Tuesday nite and the cancer patient died Wed. afternoon.

"Mary Ann, are you ready for a mercy flight?" I asked, handing her a list of medical supplies to pack for the mission. She agreed. I checked and rechecked my ambulance-red Family Cruiser, making sure it would be ready to takeoff.

I'd bought skis to change out the tundra tires for winter, and the plane was now tied on the Yukon River in front of our house. The engine was swaddled with heavy blankets to hold in heat provided by an electric heater.

The oil was indoors, ready to be warmed and fed back to the engine at a moment's notice. The river provided a solid airstrip, but the pressure ridges jutting up turned the prospect into a potential catastrophe.

"Pete, can you fire up the Cat?"

He accepted the challenge and the 12-ton D8 Cat lumbered out onto the four-foot ice, smoothing out what had been in my children's eyes an arctic expedition playground.

In the medical duplex, our kitchen radiated the aroma of baking cinnamon rolls and raisin bread. Ruby never missed an opportunity to send along a care package to missionaries; this time to the Nabingers. She'd started filling a good-sized box with a package of Jell-O, canned peas, Cheerios, cream of chicken soup, graham crackers, and a special carton of Constant Comment tea. Tucked in for their two-year-old Ralph was a jacket Mark had outgrown, a box of new crayons for seven-year-old Vivian, several grade school worksheets, and a book for their mother, Rose.

At 9:30 a.m., on the third morning, four days after the Kaltag child had gone into a coma, the ice fog subsided, although the temperature remained at minus 30°. Dressed in heavy army parkas with thick fur ruffs around our faces and other survival gear, Mary Ann and I pulled ourselves into the plane for our flight for life.

After two hours of trailing the Yukon River, we landed on the river in front of Kaltag, a village of about 150 people. We unloaded the medical supplies and drained the engine oil into a two-gallon can to be taken inside and heated. A group of Athabascan Indians crowded around us. I pulled out the Nabinger's care box and asked one of the men to make the delivery.

Quietly the rest of the soft-soled mukluked villagers escorted us to a log cabin. Death chant wails shrouded us as we entered the shadowy cabin. Nearly a dozen people accompanied the sick child's parents — all resigned to the loss of the child.

Upon examination of the 13-year-old boy, I realized the prognosis was indeed grim. Not only was the boy in a coma, but he was dehydrated with a temperature of 105°, a very rapid pulse, and one lung nearly filled with fluid. Convulsing intermittently, he made an occasional feeble effort to cough. His parents filled me with other information: he was an epileptic, and because of a recent cold resulting in bronchitis and loss of appetite, he had stopped eating and discontinued his dilantin and phenobarbital. As a result, his convulsions had increased, and he had become comatose. As if this wasn't enough, I suspected he might have active tuberculosis. Kaltag had been hard-hit by this disease.

Once the diagnosis had been formed, Mary Ann and I began emergency treatment. The death wails subsided and the 12 adults looked on curiously. Dehydration was the number one problem. Mary Ann held a flashlight and I administered an intravenous electrolyte and glucose solution followed by a glucose and water solution. There was no sterile table to assemble anything or to work from. Even washing our hands was impossible. A gastric tube was inserted for giving dilantin, phenobarbital, and massive doses of penicillin. Intramuscular injections of penicillin came next.

At the same time, we were concentrating on the critical need in this cabin, a second medical crisis was developing in another part of the village. Rose Nabinger, the missionary wife, now six months pregnant, had been spotting for several days. Her husband, Don, and daughter Vivian, were at Unalakleet (YOU-na-la-kleet) where he was speaking at the Covenant High School.

Rose described her drama in this way:

I'd awakened about 3 a.m. with definite labor pains. These increased in intensity and became closer. I began praying that God would send me medical help, "reminding" him that I was alone in that isolated village, not knowing how to contact anyone for help.

About 9 a.m. there was a knock on the door. My neighbor handed me a box from the Gaedes.

"Oh! Did the mail plane get in with all this fog?"

"No, Dr. Gaede just landed. Leonard is real sick."

I quickly bundled up Ralph and hurried out the door. I needed to go find Dr. Gaede. I was so relieved and grateful that God had sent me help.

I had to hurry, but I couldn't carry Ralph because the labor pains were increasing. The toddler struggled to hang onto my hand. When the contractions hit hard we'd both crumple onto the snowdrifts alongside the pathway. Then I thought I heard the plane revving up. I sat with frozen tears. Slowly I crawled back to the cabin in defeat.

The Nabinger family in Kaltag, Vivian, Don, Ralph, rose, April 1959

While Vivian was trying to find me, and then giving up in hope, I was finishing caring for the boy and waiting for the medication to take effect.

The moment's reprieve allowed me to shrug the tension from my neck and shoulders, and to think about other matters. I turned to Mary Ann, "Why don't you stay here while I arrange the plane seats to transport the boy. And, I'll carry the oil to start the plane." In short order, all was ready. But, for some reason, I felt I should go see the Nabingers before returning to Tanana.

I knocked on their cabin door. No one seemed to be there. I knocked again and then the door creaked open.

"Thank God! You came!" Rose startled me with her exclamation. Tears streamed down her pale face. She explained her problem and I examined her. A miscarriage was imminent.

"Just relax — I'll be right back," I calmly instructed her.

Then I uncalmly rushed out the door and ran back across the snowy trail to the first patient. Except for the boy and his parents, the house was now empty.

Mary Ann greeted me confidently, "He's stable now. We can transport him."

I wasn't sure she had ever assisted in a medical crisis under such dire circumstances, and I was relieved and grateful for her steady and dependable assistance and her positive attitude.

Gathering up my medical supplies, I whispered curtly, "Grab your parka and follow me — we've got another house call to make."

She didn't question my order. At our second emergency, Mary Ann looked around, found more wood to throw in the barrel stove, and heated water. After several hours, Rose expelled the small, dead fetus, a tiny baby girl who would never have been normal.

The crisis, however, was not over. Bleeding profusely, Rose's blood pressure dropped while her pulse rate increased. My valuable medical supplies were quickly depleted, and improvisation was imperative. I filled an empty IV bottle with boiling water, added a teaspoon of salt, cooled this solution in the snow, and administered it to my patient.

The night stretched on. I found a Native girl to spend the night with Vivian, and Mary Ann and I went back to be with the boy, who continued to show improvement.

In the morning, I returned to the Nabingers' cabin. The cabin was so cold I could see my breath in the air. The fire was out. Little Ralph had crawled in bed with his mother and was a lump beneath the blankets. Water in a glass on the table was frozen.

I restarted the fire and went outside to bring in more wood. Before long, the fire roared and water boiled in the kettle. I pulled out Ruby's cinnamon rolls and tea bags. After blessing the food, the three of us enjoyed breakfast. Rose rested easily after all her trauma, and I felt assured she'd be fine. Don would be flying in within a day and she'd be in his care.

By forenoon, the hazy sun fought off the stubborn ice fog, making our return flight to Tanana possible. I poured the tepid engine oil back into the crankcase and scraped the frost off the wings. Mary Ann and I managed to fit the groggy boy inside two army mummy sleeping bags and ease him into the plane.

The plane crunched on the hard-packed snow of the airstrip and lifted into the air. At 2,000 feet I contacted the Galena FAA and asked them to relay our flight plan and expected arrival time to the Tanana FAA.

My suspicions were confirmed — the boy had active tuberculosis and pneumonia. Within a few weeks he recovered sufficiently from his pneumonia and epilepsy to be transferred to the Anchorage Native Hospital for extended treatment.

"Well, Dr. Gaede, that was a close one, and twice as much as we'd expected," Mary Ann pondered aloud. "I wasn't sure we'd make it there in time — and then that missionary lady, wasn't that uncanny how just when she needed you, you happened to be in the village?"

I knew deep inside that this hadn't "just happened," but that I'd witnessed a miracle of God's love and power in this double feature emergency. There was a greater Physician than myself to help in these kinds of dramas.

CHAPTER 18

THE LAST ALASKA NOMADS

April 1959

CROUCHING DOWN, I moved through the darkness, feeling my way along in the low tunnel-entry to the house. Gingerly, I withdrew my hand from a husky dog's furry body. Walking in this position was nearly impossible. Three days before, I'd developed a sciatic nerve irritation and now every step was excruciating. I was more than ready to stand up straight. Finally, my host tugged at a door. A kerosene lantern illuminated a 12-by-18-foot room.

I'd finished a day of medical examinations in the Nunamiut (NOON-a-mute) Eskimo village of Anaktuvuk (An-ack-TOO-vick) Pass, which lay an equal distance of 300 miles from Fairbanks to the south and Barrow to the north. Even with my previous travels to Alaskan villages, this trip was unlike the others.

The Anaktuvuk Pass Nunamiuts were "people of the land," whereas all other Alaska Eskimos lived along the west and northern coasts of Alaska, and were "people of the sea." Until recently, the Nunamiut had moved in tandem with the migrating caribou through the Brooks Range, eating caribou, wearing caribou, and in summer living in caribou tents. The Nunamiut hadn't always been one group, but were characterized as smaller family units that crisscrossed as they moved around on the open, rolling tundra; sometimes merging, other times going their own ways. Historically, the Alaska Natives have moved about to summer fishing camps and in winter to hunting territories. In contrast to the Nunamiut clan, however, they've had a specific settlement they've returned to.

The Nunamiut depended on dogsleds for transportation, which was only possible during the winter. Given the distance to other villages, their travel was limited to intermittent treks to coastal Eskimo settlements for seal skins, seal oil, and simple food items such as tea, flour, and rice. Bush pilot Sig Wien had befriended them in the 1940s and occasionally tracked them down to trade rifles, ammunition, and a few staples for furs.

In 1949, the Nunamiut bands joined at Tulugaq Lake, and subsequently settled at what was now the village of Anaktuvuk Pass. In 1951, a small post office was established in a tent set up by Homer Mekiana. Its presence provided regular, although weather-permitting, service, making it easier for people to order and receive supplies. At about the same time, a small seasonal trading post was opened by Pat O'Connell, an Irish trapper and trader, which offered items such as coffee, tea, sugar, salt, flour, and ammunition. But until this time, Anaktuvuk Pass had been no more than a passageway and campsite.

Given their location and their nomadic lifestyle, contact with other people was unlikely. No one just happened to bump into this group. No FAA station, roadhouse, medical care, or school. No nearby fish camp, mining, or military installation to present a reason for human exchange or entice a population. No river highway or coastline to connect it with other villages by boat. No airstrip. Only small planes with floats could land on the lake in the summer and skis in the winter, thus providing a tenuous connection to the outside world. Anaktuvuk Pass *was* Alaska's most secluded village.

During my nearly two years at Tanana, I'd visited all the villages of my assigned area except Anaktuvuk Pass. This village, in the heart of the Brooks Range, and a two-hour flight north of Tanana, did not have a schoolteacher or a resident missionary to assist with radio transmission; consequently, communication filtered out sporadically via bush pilots. Besides the fact that it had been several years since a physician had visited the village, I was curious about these nomadic, inland Eskimos. I decided that an April trip up north would benefit them and me.

I would not be flying there alone.

Anna Bortel had grown to be Ruby's closest friend. Their bond and companionship got them through plenty a formidable Alaska winter night and lasted a lifetime. Anna's sense of adventure paralleled mine. She, too, was compelled to push just a bit farther into the frontier, and to tackle an experience that would take her a layer deeper into the Alaska unknown.

That winter, one of the Public Health nurses who came through Tanana had stopped at the teacherage. Recognizing Anna's interest, she revealed that a medical team would be going to Anaktuvuk in spring.

"We nurses have to fly commercially, with a charter, because of insurance reasons, but you know Dr. Gaede — he'll fly his own plane and there'd be room for you."

That set Anna into motion. She wrote the Territorial Educational Department in Juneau and requested permission to hire a substitute teacher so she could go to Anaktuvuk Pass. They wired back and told her to take administrative leave and to assess the educational situation in Anaktuvuk since there was no school there. Anna was elated.

The day before we were to leave, I was flat in bed, in spite of strong medication, with searing pain shooting down my left leg. Nothing, however, would stop me from going on this trip. The third evening I hobbled down to the Family Cruiser in the mushy springtime snow that crusted over every night. Ruby and Anna helped me pack the medical supplies for an early morning departure.

At 7 a.m. we were ready for takeoff. I told her bluntly, "Anna, I can't straighten my leg for any length of time. I'll take off, trim the plane, and tell you how to get to Bettles.

Her blue eyes clouded for a moment and then with a nervous giggle, she hesitantly replied, "Okay Doc."

The struggle to pack the plane left me exhausted, and I swallowed another pain tablet.

"If I fall asleep, or if you have any urgent questions, wake me up; otherwise, stay on this heading and altitude."

I indicated which instruments to pay attention to and what should read what. She adjusted her glasses and watched intently. The weather looked perfect and I didn't expect turbulence or any difficulty getting over the low-lying Ray Mountains.

All went well. I landed the plane at Bettles Field to refuel. Andy Anderson, a bush pilot who flew for Wien Air Alaska, and who had built the Bettles Field Roadhouse, warned us that the village of Anaktuvuk was easy to miss. He would know. He was the primary bush pilot to the village.

"You can fly right over it — even if you use the John River for a landmark and guide."

I had a lot of respect for Andy. I'd met him when he'd flown patients in and out of Tanana, and I knew he was literally a lifeline for many people in Interior Alaska, and would risk his life to save theirs. We were both the same age, but he'd logged more flying hours than I could ever imagine.

Now, the dark-haired man repositioned his cap and briefed me further, "The weather is good, but there is no airfield there. You can probably land on a nearby lake."

I always liked to know what I was getting into. Not having an official landing strip wasn't new to me, but I appreciated his instructions.

Our flight now took us north into the Brooks Range. After 70 miles, I saw the fork in the John River and continued northeast. The expansive valley varied from two to four miles wide and rose to tall, steep gray granite cliffs. Everywhere the surface was windswept clean. Trees that had clumped together were reduced to couples, then stood individually, until even those short brave survivors diminished into sporadic dwarfed willows. The stark white beauty against the blue sky contradicted the danger of the country. It seemed surreal, like a movie, or a postcard, but all in all, exactly as I'd always imagined Alaska: stunning and majestic, yet raw and treacherous.

Like a needle in a Kansas haystack, an airplane or individual person could vanish out here. In fact, last fall a Fish and Wildlife team had disappeared, and, right now a White man was lost. Certainly, no place to be forced down. I wondered how any living thing could survive on this arctic plain. But there *were* living things. A movement attracted my attention and I circled down to investigate. A large caribou herd.

"Anna! Where's my movie camera?"

If they hadn't been moving, they would have blended into the tan and white tundra. But their hooves had loosened the snow and the wind had tossed the whiteness into the air, catching my eye.

When we neared the area where I expected to find the village, I dropped altitude and studied the unchanging terrain. Sure enough, there was a collection of about 20 mounds, partly drifted with snow, all with tunnel-like entries, making them appear like the stereotypical "Eskimo igloos." On a knoll away from the other structures, a log building with an aluminum roof and crooked stove pipe stood out oddly. Since Anaktuvuk Pass was 40 miles above timberline, all wood was at a premium and had to be hauled in by dogsled. What rated such arduous effort? How many trips must it have taken to haul the logs for this building — and what was the impetus for doing so?

Anaktuvuk Pass, April 1959

I circled. It took a bit to scout out the large frozen lake a mile northeast of the village.

"The snow looks drifted and crusty," I said over the engine's roar. "Hang on."

I gritted my teeth, anticipating the jarring on my leg. After the initial impact of touchdown, the plane slid up and down along the hard drifts until I cut the engine, and it came to rest. We sat for a moment and watched as an erratic stream of parka-clad adults and children with ruddy cheeks swarmed toward us in the stiff wind. Apparently, they'd heard and seen us overhead and anticipated where we'd be landing. The Eskimos wore their nearly knee-length caribou parkas with the fur toward their body, which made for incredible warmth. The women covered their drab skin-side with a shell of colorful cotton print. This contrasted vibrantly with the monochromatic surroundings.

Anna agilely swung out of the plane, unperturbed by the powerful blast of arctic air. The prior week had been minus 40° here. At the present, it was above zero, so this was warm. It took me longer to maneuver, but finally I managed to extract myself from the plane. My plane on skis was a toy for the strong gusts that rocked and scooted it around on the ice and hard-packed drifts. Rocks or stumps for tie-downs were non-existent. I was grateful when several perceptive men went about fashioning tie-downs in the ice.

Nunamiut Eskimos at Anaktuvuk Pass in front of the Medical Quonset Clinic, April 1959

Anna and I were encircled by 20 or so Eskimos as we trundled toward an old military Quonset, set between the lake and the village. How it got there I didn't know, but apparently it had been used to hold clinic by Public Health officials before. Just that morning, the two Public Health nurses had arrived and had spent the morning in the Quonset giving tuberculosis skin tests. Polio shots were slated for the afternoon and they asked Anna to help with recording the immunizations.

A generator provided only enough heat to break the chill. Anna and I strung up blankets to ensure some privacy for the physical examinations, but it seemed *we* were more concerned about modesty than were our patients. While I began examining the 96 villagers, Anna sought out the elders in the village to learn about the educational needs of the children and the possibilities for a school.

After several hours, Pat O' Connell, the trader, invited us to his place for caribou stew. Simon Paneak, the leader of the village, joined us, as did some others. Simon Paneak had married a coastal Eskimo who had learned English from the missionaries, and now, he too, spoke fluent English — a surprise when most of the Anaktuvuk people spoke primarily their own language.

Anna plied everyone with questions; questions which were welcomed. She learned there were between 20 and 30 school age children. Simon had asked the Bureau of Indian Affairs (BIA) to provide a teacher, however, the BIA surmised that the location was too remote to build a school and regardless, no one would be willing to go there under the rigorous conditions.

"Before now they go to boarding school at Wrangell," said Simon. "The kids get hold of liquor at school. We don't want that. Anaktuvuk Pass is a dry village."

I glanced at Anna out of the corner of my eye. She was sharply focused on Simon as he spoke.

"Sit there," said Arthur (not his real name), motioning toward the caribou skins. He was a quiet man and moved slowly. After the intense and absorbing conversation over supper, we'd been invited to another villager's home. I'd made it through the entry filled with huskies and straightened up stiffly to look around. The dirt floor was covered with willow twigs and some caribou skins. I limped past the stove toward the skins and noticed a pan with dark water. He threw a handful of loose tea into the black liquid.

Without the abundance of spruce, there was no robust barrel stove for heating and cooking; instead, a puny makeshift metal stove provided tentative warmth. In spite of this, the sod house was comfortable, and Anna and I pulled off our heavy wool gloves and unzipped our parkas. I tried to find the least painful position for my leg on the pile of skins.

"Doc, isn't this a wonderful experience?" whispered Anna. Her eyes glittered with pleasure.

I nodded. My eyes roamed around the room and spotted a caribou skin mask fastened to the thick sod wall. A miniature kayak, tipped on its side on the table, lay beside an ulu with bone handle and a rounded pile of sinew string.

Arthur held out blue Melmac mugs of steaming tea, and we cautiously sipped the strong brew as it fogged up our glasses. A few other Eskimos pushed through the door. Anna's genuine interest in the people and their children showed in her inquiries and comments, and young and old responded positively to her.

"Arthur, how often do people fly into Anaktuvuk Pass?" I asked, thinking of his kind hospitality.

"Mail plane comes once a week. Presbyterian missionary fly from Barrow — fly in this afternoon, did you know?"

He let that information rest in the air and without Anna or I asking another question, he sat silently.

"It time now go to chapel." Arthur stood up abruptly and reached for his parka.

I started to drain my cup of tea, but staring into the bottom, I changed my mind. This thick mixture must have been steeping for days.

We left the coziness of the house and stepped out into a brisk wind that threatened to push our parka hoods away from our faces. Springtime sure didn't want to show her face here.

I'd learned the roughly peeled log building on the hill was the church. The recently constructed structure was the largest and most solid in the community. When the villagers had seen the design, drawn up by Rev. William Wartes, the itinerant missionary had fanned the flames of the villagers for a place to worship. They could hardly wait until there was snow on the ground to run their dog sleds the 40-some miles to timber. The church service would give me the opportunity to see this arctic wonder; however, climbing the rounded hill was a marathon of pain, buffeting wind, and unsteady footing. I didn't know how the simple cross attached to the arctic entry managed to stay erect.

For some reason, I expected to find cold metal folding chairs inside the 18- by 24-foot room, but instead there were willow boughs on the floor. People arrived and matter-of-factly sat on the floor. Anna and I turned to each other and both mouthed the word "Well?" I wasn't sure how I'd shift from standing to sitting, and neither did she. Slowly I made the transition and was able to at least lean against a wall.

The Nunamiuts enthusiastically sang church hymns in their own language, which were familiar to us. They then proceeded into a full-scale church service, complete with baptism, communion, and church membership. Inspiring and heartwarming.

The service concluded at 10:30 p.m. and as Anna and I walked out into the below-zero wind and down the hill to the Quonset hut, the bright moon was just peaking over the tall mountains in the distance. It cast a silver sheen over the village. At this moment, it was a silent night — except for the wind. I wondered what it would be like to live here long term. Would the wild beauty compensate for the rigors of the environment?

Our medical crew turned our examining rooms into sleeping quarters.

Because of my pain, I was unable to bend down and remove my boots or heavy wool pants, and awkwardly squirmed into my narrow army surplus mummy bag with all my clothes on. I didn't like this helpless feeling. I much preferred being the doctor than the patient.

The next morning, Anna completed her educational assessment. She'd collected the names and ages of children, how much schooling they'd achieved, and a projected enrollment for the near-future years. The Natives assisted her. From our conversation with Simon, she knew they shared the same goals. I knew that if anyone could make their hopes and dreams come true, that it would be her.

I completed medical examinations. The main health issues were chronic ear infections and recurrent respiratory infections. I was surprised at the small incidence of tuberculosis, which contrasted with the other villages I'd visited.

When we'd accomplished the tasks for which we'd come, a large group of adults and children escorted us to the airplane. They seemed to think nothing of the marrow-chilling wind, although at times they walked backwards to keep the wind off their faces.

Anna and I said farewells with handshakes, pats on the backs, and more smiles. A lasting impression was how good-natured these people were. Carefully, I pulled myself into the plane and adjusted the weight on my leg. The blowing parkas of our send-off crew served as multicolored windsocks, and I turned the plane toward the air stream. The plane bounced about on the uneven surface as it fought to find the sky.

Anna pressed her face against the window and waved until the people were dark dots against the white background. Then she sat in silence. I wasn't sure if she was thinking of future possibilities at Anaktuvuk Pass or just taking in the polished wonderment of the Brooks Range valley and mountains.

After a brief fueling stop at Bettles we wasted no time getting back in the air. Tanana was reported as clear, but gray clouds were settling into the Ray Mountains. I decided to avoid that bad-weather trap and instead took a valley to the west. Like a magnet, the clouds followed us as we traced the river's path beneath the low ceiling and snuck into Tanana. The village was easy to find with the evening sun reflected off the metal cabin roofs. I buzzed the village and flared out for landing. Unlike Anaktuvuk, spring *was* coming to Tanana and my river strip was deteriorating. I'd need to switch to wheels and move up to the airstrip.

We climbed out of the plane. The air felt warm, with a tinge of life. I heard a bird chirp. I suspected the pungent-smelling pussy willows were lined with gray softness.

"Tanana is so plush!" exclaimed Anna, when we climbed out of the plane. "A grocery store, school, hospital, running water and electricity, an airstrip, trucks..."

Such talk continued as we walked up the river bank to my medical duplex. I could see my children at the living room window, kneeling on the couch, looking out, and waiting for me. Anna followed me inside. Ruby hugged us both. Knowing how I usually brought back souvenirs from many of my trips, the children clamored, "What did you bring us?"

I pulled out red fox and wolverine furs, and then a caribou mask. They took turns trying on the mask and giggled.

"And wait until you see our pictures," I said. "You won't believe how the Eskimo women carry their babies. They carry them inside their parkas on their backs."

Anna elaborated. "The women can go about their tasks and the babies are warm — and comforted being with their mothers."

"Don't they just fall out?" asked Naomi.

"They put a belt around their parkas, which catches the babies under their bottoms. It makes a little seat to ride on," I explained.

"Yes, and guess how they get the babies out?" laughed Anna. "They don't just undo the belt and let the babies drop out. No, the mothers lean over with their heads nearly to their knees and the babies come out the top of the parkas where their mothers gently catch them."

Anna tried to demonstrate and nearly toppled over. Everyone laughed.

"The little tyke came out into the wind and cold. I felt so bad for him. Of course, I asked her to put the baby away. She pulled her parka up, pushed him around to her back, bent over until he slid up her back, and then belted him in."

I looked around. A green carpet, rather than twigs and branches, covered the floor. There were no caribou skins to sit on, but chairs accompanied the kitchen table.

"It's hard to believe that such a different world is only hours away."

I walked into the living room and eased onto the cushy sofa.

"Naomi and Ruth, could you help untie my bunny boots? I've had them on for three days."

At the time, I didn't expect I would return to Anaktuvuk Pass and discover that a baby in the Hugo family had been named for me. Not only would he have my first name of "Elmer," but a middle name of "Gaede." Neither did I know that within a year Anna would return to Anaktuvuk Pass to establish a school. She'd have even more stories to tell — enough to fill a book of her own.

I did know that the reception of the Anaktuvuk Pass people was a special and treasured gift. I hoped I'd returned gifts of health and longevity.

More about Anaktuvuk Pass

In the spring of 1962, the Gaede family, minus Mishal, flew up to see Anna Bortel. She lived in a sod house and was offering the Nunamiuts what they had dreamed of: education for their children, as well as an opportunity for the adults to learn spoken and written English. Naomi, Ruth, and Mark were the first White children many of the Eskimos had ever seen. This story can be found in "From Kansas Wheat Fields to Alaska Tundra: A Mennonite Family Finds Home." Anna's stories are told in "'A' is for Anaktuvuk: Teacher to the Nunamiut Eskimos."

CHAPTER 19

2,000 FEET ABOVE THE YUKON RIVER

May 1959

"DOC, WE HAVE A BIG PROBLEM. Do you think you could fly down?"

On this particular afternoon, I was talking over the radio with Russ Arnold, Arctic Missions missionary at the village of Ruby. I never knew what I'd hear when I opened on the air with "This is KIK 731, Tanana. Standing by for medical traffic."

Russ clued me in fast.

"This woman has really gone berserk. Smashing furniture, screaming, threatening to kill anyone who comes near her. We've tried to calm her down, but nothing's working. We don't know what to do. She's going to hurt herself — or someone else. Can you help us?"

Russ was not one to be upset easily. The villagers looked to him in times of crisis — and this was one of those times.

Russ and Frieda Arnold family at the village of Ruby

Glancing out the window, I replied, "Okay Russ, the weather looks good. I'll try to be there in an hour-and-a-half."

Summer house calls with the 24-hour daylight and moderate temperatures were usually as simple as making hospital rounds.

As a bush doctor, the unexpected was to be expected; yet, just when I thought I'd seen everything, something would astonish, and even unnerve me. The previous winter I'd been awakened in the middle of the night by "Doctor, doctor, Lucy's awfully sick." I couldn't believe my ears. The voice seemed right in my bedroom. Was I dreaming? I reached for my glasses and swung my legs over the side of the bed. I walked into the kitchen and bumped into an older White man! I recognized the white-haired gentleman as Lewis Kalloch. One shoulder strap of his faded time-worn overalls had come undone and his dark-framed glasses were askew. He gently, but persistently repeated his concern. I suggested he sit a moment while I got dressed.

Like many others, Lewis had journeyed to Alaska in the early 1900s in search for gold. When he found enough to afford a wife, he married Lucy, an Athabascan Indian. In summer, they would sit across the road from their two-story clapboard house, on a backless log and plank bench in front of the river and quietly watch boats travel on the waterway. Their fondness for each other was evident and endearing.

I was used to middle of the night calls from the hospital, but this caught me off guard. Now, with no medical bag, and only my stethoscope, which seemed to trail me everywhere, I left the house and followed him down the road to his house, on the other side of the school. Nothing stirred in the village. No noise of chainsaws or barking dogs.

I examined Lucy carefully. She had pleurisy. Lewis and I talked about what to do to make her comfortable. His tender concern for her was reflected in his questions. After I felt the couple was assured, I left the old-timers and walked back home in the nighttime dusk and 34°, feeling warm inside. The experience left me shaking my head and smiling all at once. I'd never had someone make a doctor's appointment quite like that. Neither had I received the kind of payment that came a few days later.

Ruby happened to be walking back from visiting Margie Gronning, when the old gent stopped her. "You are the doctor's wife?"

She nodded.

"How is your potato supply?"

Ruby told him that the only ones she had were frozen. Shortly after she arrived home, he showed up on our doorstep, this time *knocking* on the door. With dignity he presented her with a peck of potatoes he'd grown the summer before.

"The doctor is so good to me."

The feeling was mutual.

This incident made me think of my boyhood. Our family had been so very poor and money scarce that we, too, bartered for goods and services. Harold and I would ride along with our father in a truck loaded with wheat, and at the grain elevator in town he'd exchange it for flour. Mother would greet us at the doorway, her calico apron buttoned in the back and her sleeves rolled up, ready to bake the very basic of life — bread. Flour came in colorful coarse cloth sacks. Our parents would always choose two bags with the same design since two would make an everyday or school dress for my sister, Lillian; otherwise, the fabric was cut and sewn into aprons, towels, pillowcases — and even underwear. It wasn't the easiest fabric to work with, or to wear, and ironing only minimized its scratchy feeling. The companies that made the flour and feed sacks figured out this purchasing incentive and created new patterns.

When our mulberry tree was ripe, Harold and I would shake the mulberries to the ground. Then we'd stomp the red-purple berries with our bare feet to press out the seeds, which we'd then wash. Once dried and packaged, we would accompany our parents to the grocery store where the "bird seed" would be exchanged for staples such as sugar, salt, baking powder, and a ring of baloney. Oh, how we boys loved baloney! This was about all we needed with the provisions of the farm: milk, eggs, fresh fruits and vegetables, chicken fryers and sausage.

To glean a few more dollars, Dad would take his two horses and a plow to town to cultivate people's gardens. He'd get $5.00 for his services, which was a fortune to us! On other occasions, he'd mow the sides of the highway with a horse-drawn mower and get cash from the state highway department.

Our mother would barter with the Watkins man who would drive from place to place with a van full of spices and flavorings. She would trade a hen or two for flavorings of almond, maple, and vanilla, and spices such as cream of tartar. The maple would be added to powdered sugar frosting and spread on "long john" doughnuts, which were cut in rectangles from rolled out sweet dough, and deep-fried, rather than turned into the usual dinner and cinnamon rolls.

The people here bartered and reciprocated, in comparable ways, by using their proficiency with animal skins, hunting, fishing, woodworking, and native grasses. I'd received pieces of ivory, woven baskets, bead-worked footwear, and other Native handmade items. Ruby and I were touched by Lewis's gratitude; yet after that middle-of-the-night appointment, my vigilant wife did a double-check on the door locks before we went to bed.

The situation down river was critical. I tracked down Mary Ann. Even though Alaska village medical practices were new to her, she never left my side or let me down. I described the situation at the village of Ruby.

"I want you to come with me. You, as a woman, will be more able than me to calm down this patient and persuade her to take medication." From my experience of working with distraught patients, I'd observed this effect.

"I imagine you'd like me to escort her back on the charter plane, too," said Mary Ann, reading my thoughts.

"Precisely."

Mary Ann prepared medical supplies and I called to see if a charter would be available to fly the patient back to Tanana. We loaded my red pride and joy and within minutes, my tundra tires stirred up a dusty trail, left the airstrip, and rotated to a stop as I gained altitude.

I'd flown up and down the Yukon many times, yet my eyes always scrutinized the landscape below for wildlife, as well as for potential emergency landing spots. Now, I watched a kicker boat pushing against the current, leaving a momentary mark on the river. Mary Ann pointed out a bear catching fish at the mouth of a small incoming river. Within an hour, we'd covered the 110 miles to Ruby, circled, and landed on the village strip, which was a rolling hill with a crest in the center.

The village was named after red-colored rock found along the banks. It had seen its day in the early 1900s when prospectors struck gold at nearby Ruby Creek and Long Creek. At one time, 1,000 miners lived in the village and along neighboring creeks. Although, then, it was primarily a non-Native village, the Athabascan had found work at sawmills and other businesses, as well as a market for fish, sleds, snowshoes, firewood, sewing, and furs. After the brief era of the Gold Rush, the population decreased dramatically, with the remaining population primarily Native. Now, depending on the time of year, 60 to 110 people lived in the village.

Russ was standing at the edge of the strip. "So glad you're here, Doc."

He hastily guided us through a group of curious villagers to Roberta's cabin. (Not her real name.) From the outside, it was a typical cabin with a gas-powered washing machine sitting on the front porch. I anticipated pandemonium as we neared, and was surprised to hear nothing. We stood cautiously in the open doorway. I turned to Russ and he motioned for Mary Ann and me to go inside.

We stepped forward tentatively — uncertain about the greeting we might receive. My eyes adjusted to the darkness and I found a middle-aged woman slouched on a shelf-like sleeping bench against the far wall of the one-room cabin. She sat sullen and disheveled. I slowly introduced myself and Mary Ann, and then tried to start a non-threatening conversation about her frustrations and anger. Inch by inch I moved toward her. A red vinyl-covered kitchen chair served as a barrier between us, so I pulled it to one side and sat down. Mary Ann stood behind me.

"Wouldn't you like to get away from your home and family for a few days and rest in the hospital?" I asked softly.

She made no verbal response, but pulled back, tucking her legs beneath her, and pressing her body against the wall.

"We're here to help you," said Mary Ann.

No reaction.

I shifted my attention from Roberta to the broken white dishes on the wide-planked floor, dirty clothes strewn around the room, and a wooden chair smashed against the cold barrel stove. I was concerned that she would erupt into another rage and wanted to give her tranquilizers.

I tried again. "Roberta, I'd like you to..."

"No!" she screamed. She jumped off the bench and stared wildly, poised for flight out of the cabin.

I could understand Russ's helpless feeling. I leaned my body toward her path of escape. Mary Ann resumed talking in placating tones. After what seemed as long as an Alaskan summer day, she consented to go to the Tanana hospital.

I stood up feeling as though we were making headway. "Roberta, I've arranged for a plane to take you to the Tanana hospital. You just need to pack a suitcase and we'll go down to the river where the charter will come in on floats."

I'd assumed too much.

Roberta scowled at us and screamed, "I won't go anywhere unless it's with you, Doctor!"

"No, Roberta, that will not be possible." I felt as if I were dealing with a child. "I will meet you in Tanana when you fly in on the charter."

"No! No! No!" she shrieked, stamped her feet and shook her head until her frayed braids struck her in the face.

I expected she'd start throwing things or run out of the room, and braced myself, but instead, she dissolved into a sobbing heap on the filthy floor.

"I'll be good. I'll be good. Don't leave me."

I motioned to Mary Ann to follow me and walked outside into the sunshine. Taking a deep breath of the fresh air, we considered options.

What would happen if we couldn't get her on a charter? What would happen if she were to fly back with us? What would this disturbed woman do in the confines of a small plane? What were the liabilities?

After weighing the alternatives, I decided to go ahead with an emergency flight. We did, however, need to map out our strategy for any in-flight crises. "Roberta will sit in front with me, and you will sit behind her with a filled syringe of a hypnotic drug, which you'll use if she becomes agitated."

When we reentered the cabin, Roberta stood waiting meekly with her small suitcase in hand. Russ and several Native men followed behind as Mary Ann and I kept Roberta between us in our march toward the plane. The three of us found our designated seats. Roberta docilely buckled up, and with a sigh of resignation, folded her hands in her lap.

The plane rapidly climbed into the calm blue, cloud-smudged sky, and about five miles out of Ruby we were at 2,000 feet above the Yukon River. Without warning, our compliant passenger exploded. Grabbing the plane door handle, she frantically shoved it open and with a desperate lunge thrust her foot out into the swift air.

"I'm going home!" she screamed.

Instinctively, I rolled the plane over in a steep left bank to keep my unwilling passenger from falling out. Mary Ann locked her arms around Roberta's neck as Roberta clawed at the door in an attempt to regain her balance and continue her escape. The compass rotated wildly, the airspeed indicator crept down, and the turn-and-bank indicator slammed to an almost vertical position. We roller-coastered in the air.

Then, with the 60-degree angle of the plane, the weight of the door created enough discomfort that Roberta retracted her foot. I leveled out the plane, which caused the door to slap open and shut. The plane circled erratically in the untroubled sky. After a few long moments, Roberta appeared to be settling in. I cautiously reached across the woman to latch the door. Mary Ann released her arm-hook. I took a deep breath, and simultaneously tried to steady the plane, be on guard for any sudden movements, and furtively tell Mary Ann to prepare to administer the injection.

Roberta did not remain composed. Again, she attempted to make an emergency exit *home*. I instantly shifted my position and with my left hand on the stick, firmly hooked my right arm over one shoulder and under her other.

"Give her the injection!" I yelled at Mary Ann, who was trying to maintain *her* balance.

As if maneuvering in the cramped, pitching cabin, wasn't impossible enough, Roberta was thrashing about, her legs kicking, and arms flailing. After a seemingly endless struggle, the needle found its target. Following some tense minutes, our patient relaxed. Apprehensively, I released my grip, and again, secured the plane door. Neither Mary Ann nor I felt relaxed or secure. For the last 100 miles, we flew in silence, not wanting to excite our passenger in any way.

Tanana FAA had notified the hospital of our estimated time of arrival and the ambulance was waiting when we touched down. Roberta was initially treated at the Tanana hospital, and then transferred to the Alaska Psychiatric Hospital in Anchorage.

"Well, Dr. Gaede, mixing your medical duties with flying certainly brought on some excitement today," mused Mary Ann after we resumed our usual duties.

"Yes, the kind I like best when there is a safe ending. Now what *was* on today's schedule?" I asked jokingly.

Rapid footsteps echoed down the hallway.

"Dr. Gaede," the emergency room nurse tried to catch her breath. "Mrs. Gronning was just admitted. She's miscarrying and hemorrhaging."

This was no time to relax. I'd made it back just in time.

CHAPTER 20

THE TOP OF ALASKA

July 1959

I FELT RESTLESS. My book of Alaskan adventures had empty pages. Yes, I'd covered a lot of country, but I wanted to touch the top of Alaska.

Another force drove me. After four years in the land that had stretched every fiber of my being, challenged me physically and mentally, thrown me to the edges of survival, and given me more than I'd ever dreamed of, I was being transferred "Outside," to the lower 48. Now as I stood on the banks of the commanding Yukon River, watching a bush plane circle overhead, the call of the wild rang loudly in my ears.

I could make this final expedition on my own, but the camaraderie of a kindred spirit would add to the pleasure, and to the safety. I knew Ruby would be slightly less anxious if I took someone along, just in case I was forced down, not that I expected such trouble. I walked up rain-sprinkled Front Street, past buildings that had become well-known to me. "Hellos" were exchanged for "Hi, Docs." After the strange village welcome, I was no longer a stranger. I reached the Arctic Mission's house-chapel door and tapped on the door. A tall man, near my age, answered.

"Mel, how would you like to go to Barrow with me?"

Mel Jensen, his wife, Pat, and two young children had recently moved from Nenana and were replacing the Gronning family. I missed Roy, the big Scandinavian with boyish enthusiasm for life and the strength to meet any adversity in this wilderness. However, this newcomer was turning out to be very approachable.

"What exactly do you have in mind?"

Uncannily, it appeared Mel was ready to not only fill the missionary role, but the buddy-role that Roy had had with me.

"Sounds great, Doc," he said without hesitation.

Now, there would be two sets of eyes to detect landmarks and two minds to generate options.

I'd set the date of July 8 for our trek to the northernmost point of Alaska. Then I mapped out the 1,300-mile trip above the Arctic Circle, through the Anaktuvuk Pass in the Brooks Range, to Barrow, and then back home along the coastline.

In contrast to the stingy-sized J-3, which I'd taken on previous flights up north, my new Piper PA-14 opened its doors to our needs. After searching for fuel at destinations on previous trips, I packed in 20 gallons of extra aviation fuel. We added over 100 pounds of emergency gear, including clothes, food, guns, and repair equipment. Chocolate Hershey bars had become part of my emergency gear, although they were seldom consumed — or replaced — and therefore remained stashed in the back of the plane, freezing in the winter, thawing in the summer. This repeated cycle eventually turned the brown bars white and the flavor faded, thus reducing their appeal.

The week prior to our departure, Alaska showed her worst side, with high winds, electrical storms, and frequent rain squalls. The Yukon River beat against the shore, and trees fell over in the woods and on the riverbanks. The unstable ceiling vacillated capriciously. Too risky for sane flying. Snow showers taunted the upper coast of Alaska — even though it was mid-summer. On the evening before July 8, the daily FAA weather sequences gave no hope of improvement. Despite the gloomy forecast, I went to bed planning to carry through with my scheduled departure.

My optimism was rewarded. By 10 a.m. the low morning scud broke up and the ceiling lifted to 2,500 feet. The barometer crept up to nearly 30.00 inches and the Kotzebue coastal area showed clearing with westerly winds. The improvements spread and by 11 a.m. Anaktuvuk Pass was reported open.

I sprinted to Mel's house. He was vigorously digging up his small front yard. When he saw me, he stopped and leaned on his shovel. "Thought I'd make some order in here. Plant some grass for a lawn and put in a garden." Sweat gleamed on his vastly receding hairline.

From what I'd seen, Mel was definitely an order and organized type guy. Nothing random. I liked that.

"Ready for a trip north?" I grinned. "The weather is the best it's been in ten days."

"I'll be right over!" He moved quickly, gathering up his tools and disappearing into the house.

The plane left the ground at 12:05 p.m. and we were off to see the top of Alaska. We skimmed under a broken cloud deck and slipped through a pass at 3,500 feet in the towering Ray Mountains. On the other side we burst out onto the flat, north side which was filled with small lakes. Light rain showers accosted us and persisted to dog us as we flew over Bettles and northward toward the Brooks Range. Conditions didn't improve. Stringy rain-threatening gray clouds hung like a curtain before us. I radioed Bettles Field for a pilot report. No local pilot reports. No one had made it through. Bettles Field FAA, however, reported favorable conditions farther on at Umiat (OO-me-at) and Barrow. I replied that we would push through.

As we advanced toward the pass, we encountered a strong southerly wind funneling above the turquoise-green John River. Paradoxically this was actually encouraging, since it cleared an open airway for us to tunnel through. To my surprise — and relief — despite the intermittent drizzle, which beaded up on our front windows and streaked back along the sides, the ceiling held at 1,000 feet. As we neared the pass, the skies gradually lifted. This good news lasted only minutes before the pass turned on us with violent 60 to 70 mph winds. The plane was tossed about in the unstable air. I pulled it into a climb and we settled down at 5,000. Unruffled, blue sky replaced the retreating darkness.

Having successfully negotiated the pass, we followed a compass heading toward Umiat. The topography kept changing as we moved northward. No more mountains. Turquoise mottled lakes spotted the utterly flat terrain and looked like huge cut and polished rock halves. Narrow, rust-streaked rivers zigzagged around the flat, treeless tundra, and cut it into pieces of spongy patchwork. The geography copied itself over and over, making checkpoints difficult to identify. Eventually we saw the blue Colville River, with its huge mile-long sandbars, which reassured us of our whereabouts. Umiat lay dead ahead. In 1944, the U.S. Navy Seabees had drilled the first oil well at Umiat for the former Naval Petroleum Reserve. The site later became a U.S. Air Force Station.

The now abandoned military field had two Wien Airline radio men stationed to assist the airline's planes. We landed to refuel with our own gasoline. We'd flown for three hours, prodded along by a stout 40 mph tailwind.

The Barrow compass heading was 295° but the strong crosswind made me correct with a 260° heading. I hoped I'd find a landmark. We were on our last leg and now flying over a myriad of nameless semi-frozen lakes and braided serpentine rivers. The progression of water to ice in the lake created a three-dimensional look with shades varying from pale aqua to teal blue, to blue-black and green-black. Each year these old lakes struggled to thaw, but never won.

The barometer was dropping gradually and fluffy scattered clouds nudged us down to 2,000 feet. After flying for an hour, we spotted a checkpoint to our north, Teshekpuk (Ta-SHEK-pook) Lake on the edge of the Arctic Ocean. We were on course, but the ceiling kept creeping lower and we dropped altitude again. I felt uneasy flying this low over open water and deviated to the south, then flew on the edge of Dease Inlet. Our fascination had been downwards, at huge caribou herds; when we re-focused on the horizon, we noticed a white bank of fog or low clouds.

"What do you think that is?" I asked Mel.

He was as puzzled as I. We drew closer. The flat earth and sky blended together with little relief, making dimensional perception difficult. When I finally figured it out, I was awestruck. "Oh! Look! It's a tremendous ice pack shoved up against the shore!"

Mel had scanned more than the shoreline.

"Look! There's Barrow right ahead of us," he said excitedly.

We'd made it! The top of the world sprawled out about a mile along the shoreline.

Since there was only a military airport where pilots needed prior permission to land, I scoured the beach for a suitable landing spot. Boats, dead walrus, and ice chunks obstructed our proposed runway. As an alternative, I chose a deeply rutted road in the fine sand at the Browerville end of Barrow. Lowering the flaps, I settled down carefully with the large tundra tires riding over the ruts.

Within minutes, beach-combing Inupiat Eskimo children curiously and laughingly surrounded the plane. So, this was Barrow. I was eager to go exploring, and follow these young, ebullient hosts, but the wind tugged at my plane and we needed to tie down. Even though it wasn't a cold wind, it was only 40°, which was typical for the summer here, and the plane rocked to-and-fro. Glancing around, the only possibilities we found were a "honey" barrel (containers used for human excrement since plumbing was not possible in the frozen ground) and two small barrels of sand.

With this task accomplished, we walked toward Barrow. Besides being the farthest point north, it had the largest Eskimo settlement in North America, with a population of 800 to 1,000. Our merry hosts had hung around and now we asked them for directions to Miss Felkirchner who the previous year had been the Tanana Day School principal.

When Florence opened the door, the typically staid and formal matron smothered me in surprise and joy. Without hesitation she invited us in, put on the tea kettle, and unreservedly bombarded me with questions about Tanana, Anna Bortel, and her former students. I felt like an audio newspaper. The conversation was informative to Mel, too, since he was newcomer to Tanana and eager to learn as much as he could. I'd never seen her so happily animated and expected her hair would at any moment free itself from the bun on her neck. Barrow was not like Tanana, and this woman struggled with the bleakness: lack of vegetation, gray gravel, muted sand, and overcast skies — both in summer and winter.

Florence offered us space in the school to throw down our sleeping bags, and then invited us to join her for a caribou roast supper. Before we pushed our well-fed stomachs away from the table, she let us know when breakfast would be served in the morning.

At this top of the world, the sun had finally risen on May 10 and would not set until August 2, so walking around after supper allowed unlimited daylight for sightseeing. (In reverse, the sun would go completely down on November 18 and not even skim the horizon until January 24.) We were quick to see how the permafrost, ground that remains frozen for two or more years, which underlies the entire Arctic region of Alaska, affected life in Barrow. For one thing, the frozen ground was nearly impossible to dig into. Mel and I kept finding clusters of six to ten "honey" barrels, and questioned some passersby. We learned they were common and used by everyone, including the Public Health Service hospital and Bureau of Indian Affairs grade school. They would be around the village until winter when they could be hauled by dogsled out onto the tundra.

Obtaining water was an ordeal as well. The precious commodity was hauled in from freshwater lakes in the form of liquid or ice. The cemetery exposed yet another problem. Complete burial was futile and some of the caskets lay above the ground with a few strips of tundra draped over them. Where the unpredictable permafrost had changed positions, the caskets had gradually and precariously sloped.

In other ways, the permafrost was actually useful. For example, it created a solid foundation for houses. Ironically heat on the permafrost would interfere with this sturdy base. Many of the houses stood on four- to five-foot pilings, thus preventing the heat in the house to thaw the ground beneath them and tilt the structure awkwardly on a semi-thawed foundation.

Gardening was obviously impossible, although farther from the shoreline tundra flowers showed their faces in short treats of pink and yellow.

We constantly dodged the permanent gray puddles which formed on the gravel roads and paths. Drainage was a never-ending problem.

None of these challenges, however, deterred people from outside the village to fly to the top of the Alaska world.

"Looks like this place is really geared toward tourists," observed Mel.

I agreed. Nowhere else, with the exception of Anchorage, had I seen the local people clamoring for the attention of visitors. Everywhere we turned there were tours, souvenirs, and demonstrations. Even though I didn't mind another opportunity to experience Native traditions and festivities, this seemed contrived.

Wien Airline, which had an agreement with the military airstrip, had daily tourist flights connecting with Fairbanks, and one of Barrow's businessmen appeared to have a monopoly on the hotels and restaurants. Since Mel and I had flown in on a private plane and stayed outside the customary lodging places, we were often given the cold shoulder. For instance, no one would give us information about sightseeing or where to get airplane gas. Sightseeing was then something we just stumbled upon.

We discovered that every evening during the tourist season a group of Eskimos displayed their traditions by demonstrating blanket tosses, performing Eskimo dances, and offering dogsled rides. The "blanket" was made of a round 10- to 14-foot diameter walrus skin and functioned like a trampoline. We crowded in with the other spectators. Ten to fifteen adults and children gripped the edges of the skin, all dressed in parkas, with the women wearing multi-hued head scarves knotted beneath their chins. One individual jumped in the middle.

On this occasion, a lovely young woman in a scarlet parka with intricate trim showed off her skill. She warmed up with short jumps, making the audience wonder how high she could ascend. The cheering revved up as she smiled and leaped higher and higher, kicking her legs as if climbing into the air. Even though all eyes were on her, this was a team sport. The second she hit the blanket and prepared to spring up again, her team pulled back, which aided her bounce back into the air. Up she went, her white fur parka ruff blowing around her head and shoulders. Down she plummeted, her black ponytail flying up above her.

After a while, we concluded that a medium weight person, with the help of tossers, could jump 20 feet, and that the jumper was allowed to continue until he or she lost their balance. The blanket toss was said to originate when Eskimo hunters used this method to spot whales or seals in the distance. At this time in history, it was a game, and a part of every Eskimo festival or competition.

Dogsleds rides were another big hit with the tourists. Even though it was not uncommon for Barrow to have snow showers in the summer, at this moment there was no snow on the ground. Undeterred, the Natives made the most of their seasonal enterprise, and had ingeniously fastened small wheels to the dogsled runners. The grinning huskies pulled the ecstatic sightseers around the village. Everyone was happy. Neither Mel nor I felt compelled to join in. From my experience, this was definitely not a real-life experience. The exotic appeal had disappeared for me at Lime Village, leaving me with enormous admiration and high regard for the Natives who depended on this mode of transportation for their livelihood.

It had been a long day. Flying here had required mental vigilance. The still-shining sun showed no intention of bedding down, but we meandered to the school and crawled into our sleeping bags.

About 5 a.m., shrill high-pitched screeching wind woke me up. It sounded like a storm working itself into a fury. A chill went down my back as I thought of my plane anchored near the water's edge, exposed to the full force of the wind — and water.

I shook Mel roughly, "Get dressed. We've got to get to the plane!"

I opened the school door and blinding sand filled my eyes and pelted my face. Leaning into the 50 mph tempest we pushed down the road. My red cap didn't provide any protection. I stuffed it in my pocket before it blew off my head. The ice pack had moved out and ice cakes thrashed around in the tumbling ocean.

"How are we going to get through, Doc?" Mel shouted.

In front of us 40 to 50 feet of ice-choked water covered the road between us and the Family Cruiser. I squinted against the gritty wind, and my eyes followed the length of the water. My heart thudded loudly in my ears as I realized that the water had become a channel across the road, cutting off my rescue attempts. I imagined my red aircraft torn by ice and wind, and either submerged in salt water or littered across the beach.

"Mel, I can't give up!" I yelled hoarsely. "I'm going to walk around."

"Doc, there's no way you can do that!" he retorted, grabbing at my arm.

I pulled away, tucked down my head down, and started along the water's edge. Nature fought me. My tall rubber boots sank into the wet sand and the wind tore at my near knee-length army coat. When I looked up for a moment to get my bearings, I saw ahead of me an elevated four-inch pipeline, bridging the channel of water to dry land. Shakily, I straddled the horizontal pipe and inched above the ice-filled turbulent water. Once on the other side, I attempted to run against the unrelenting wind.

Through the gray squall, I saw the airplane was upright and untouched by the frothing water, even though one of my tie-downs had torn loose, allowing the tail to swing out toward the ocean. The broad wings bounced about, but the tires remained rooted securely in the sand. I quartered the aircraft back into the wind and searched for more tie-downs. Some large chunks of heavy scrap iron had been carried in with the storm and made excellent anchors.

About the time I secured my plane, several Inupiat men and women plodded out to check their skin boats, which were tied along the water's edge. The high breakers, loaded with enormous icebergs, had covered and damaged several craft. No words were needed — or possible. We all recognized what needed to be done. I labored with them to salvage their boats and pull them to safety. Once accomplished, the small assembly disappeared into the raging mist. I made my way back the same way I'd come and found Mel standing right where I'd left him. His parka ruff sucked tightly around his head so that his eyes were barely visible

"Doc! Are you okay?"

"The plane's okay." I panted, trying to catch my breath.

By this time, we were on schedule for breakfast. I felt as though I'd put in a day's work when I stumbled into Miss Felkirchner's peaceful kitchen.

"Looks as though you could use some hot breakfast," she said cheerily, flipping pancakes and pouring me a cup of steaming tea. After removing my wet, sandy boots and wiping off my glasses, I slid into a kitchen chair.

After breakfast I sought out the weather forecast. Getting socked in here was not my plan. The courteous crew at the U.S. weather observation station provided detailed information, although none was encouraging. The storm, the worst in eight months, was forecast to remain at least a couple of days. We wouldn't be taking off any time soon. We went back outside under the sagging gray sky. At least now that I knew my plane was secure, the wind didn't seem as fierce.

Children or adults followed us around so we were never without directions or guides — not that we could have gotten lost in this level, one-road, sparsely built village. We decided to check out a general store to see what it might offer. We found tourist trinkets and common necessities. I bought a postcard with a polar bear on the front to send to my brother, Harold. The White man running the business suspected we were new to the area and wouldn't let us leave until he'd told us about Barrow's history. The wizened man with hair as frosty as the world he lived in, drew himself up as tall as possible, like he was a guest speaker at some convention.

"You know, boys, Barrow was named for Sir John Barrow of the British Admiralty by Captain Beechey of the Royal Navy in 1825. Yes, siree, that Captain Beechey was plotting the Arctic coastline of North America."

He stopped to catch his breath. "By the way, fellas, do you know the name of his ship?"

He didn't pause to let us even guess.

"HMS Blossom. Funny name for a ship up here, don't you think?"

We nodded.

"That was just the start of things. Pretty soon, whalers showed up — you know those fancy ladies down in the states wore them corsets."

He raised his eyebrows and then winked.

"Full of stays made out of whalebone — and whalers could get $5.00 a pound."

He sounded like a history book and I wondered if he was truly enthralled by the subject matter or if he'd just gotten used to repeating it dramatically for spellbound tourist groups.

I believe he could have spent the day talking to us, but finally after interrupting him with a "Thanks so much" and a friendly pat on the back, we continued our exploration.

Of course, I wanted to visit the Public Health Service hospital. After meeting the Medical Officer in Charge and getting a tour, we stopped at Rev. John Chamber's house. One fact about Alaska villages was that no one really needed an invitation to drop in. With such small populations and limited communication beyond their community, residents viewed guests as fresh conversation, news bearers, and social opportunities. John Chambers was referred to as the Presbyterian "Flying Missionary of the North." His home base was Barrow, but this pastor-pilot's parish was 40,000 square miles of tundra, including the village of Anaktuvuk Pass. Mel and I leaned forward in our seats as he described his rewarding friendship and engrossing work with the Eskimos at the various villages.

The next day, the winds had lessened to 25 to 30 mph, and the clouds lifted to 600 feet. A few local planes pushed against this upper limit, and we were told that we couldn't expect any better conditions. All the same, weather wasn't our biggest restraint. Obtaining aviation fuel seemed to be a non-ending challenge.

Since local pilots had their gas supply shipped in by boat, no store stocked aviation gas. Outsiders rarely flew in, so there was no need for this type of fuel as a commodity. At wit's end, I finally tracked down a pilot working for the weather station who suggested I try to get gas from Mr.

Brower, who owned most of the businesses, and was part of the founding family of adjacent Browerville. Mr. Brower was getting on in years and was confined to his room. We were not received well, and it was very frustrating talking to him. I finally resorted to name-dropping, mentioning Mrs. Felkirchner, Rev. Chambers, and Public Health Service personnel.

Eventually he tired of our begging and agreed to sell us one barrel of gas — if we could find a full barrel somewhere near his house. We searched, found a 55-gallon barrel, and paid him $1.25 per gallon. By now the storm-carved inlet that had separated us from the plane had retreated and we caught a ride to fill our gas cans. With that feat behind us, we prepared to resume our journey.

The plane bounced lightly over the beach ruts and was airborne within a couple hundred feet. We started down the coastline from the Chukchi (CHUCH-chee) Sea toward the Bering Sea, which was an easy route to navigate with the obvious landmarks of the shoreline. I filed a flight plan to Kotzebue with Wien Airlines.

When we took off, low scud prevented me from gaining much altitude, and I hovered at about 100 feet as I pointed the plane along the coastline. Eventually I would arrive at Point Hope and then retrace my route back from the polar bear hunt. Meanwhile, to allow for the 30 mph crosswinds, I had to crab the plane, pointing its nose toward the wind, which gave the illusion of flying sideways. The low altitude was scenic, but unnerving.

Several miles south of Barrow, we flew over the Will Rogers Monument, which commemorates the 1935 death of the American humorist and pioneer world-circuiting pilot. Rogers had landed there, seeking directions to Barrow. Upon takeoff, his plane rose 50 feet in the air, stalled, then plunged into the river below. The day before, I'd been offered $50 to fly a tourist to this monument. In yesterday's storm it would have been suicide.

After an hour and a half of following the sharp coastline, we came to Wainwright. I wanted to visit the Wycliffe Bible translators there, so

I checked out the narrow beach for landing. I was hampered by a stiff crosswind, but kept the plane in control and touched down without a problem. Everything seemed fine until I turned the plane toward the low shore banks. In spite of the wide tundra tires, the plane bogged down in the muddy sand. From the air, I hadn't seen this slough, which was formed by water runoff from the village.

"There's always a surprise with these landings," said Mel, taking it all in stride.

We climbed out and each put our weight against a strut. As had become typical, and much appreciated, a group of Native men hastened to our aid, and inch by inch we tugged the plane farther up the shoreline.

The uniqueness of the village made the stop, and the landing predicament, well worth our time. Here at Wainwright, the tundra was so soggy that the villagers had constructed "sidewalks" throughout the village. I'd never seen the like. The sidewalks consisted of 55-gallon gas drums, turned on their sides, placed next to one another, and fastened together. On top, planks provided a flat surface for walking. This ingenious solution was captured by my 16 mm documentary. After Naomi and Ruth viewed it, they thought they'd like to move there. I could imagine them spending entire days and weeks running and running over the barrels.

After an hour's worth of conversation and tea with the translator couple, we felt refreshed and ready to move on. The Cruiser pulled free of the mire and into the tranquil sky. Weather conditions steadily improved. No crosswind. Clear sky. High altitude. Flying seemed too easy. Mel and I kept motioning the other to look at something below or on the horizon. I relaxed and enjoyed each moment while we steadily moved past landmarks. Point Lay. Cape Beaufort.

Then it happened.

"What's that white stuff?" I asked, sensing danger, but not knowing exactly why.

A solid wall of white crept toward us from the ocean, shutting down our forward visibility.

"Fog? Low clouds?"

We approached the edge of the eerie wall. In terror, I recognized the whiteness.

"Blizzard!"

Immediately, I executed a 180-degree turn, pointed the nose down, and, breaking out in a sweat, searched wildly for an emergency landing spot. My mind snapped out a picture of a mounded sod hut along the beach, about 12 miles back. I opened the throttle to race the storm. We approached our possible refuge with the storm ready to engulf us. There was no time to do a preliminary drag of the narrow shale beach. I set the plane down — hoping for the best. The landing went well, but we weren't home free. I maintained power and taxied nearer the sod cabin; only then did I cut the throttle. We climbed out. Within seconds, the storm struck, and the gale whipped the breakers over our plane tracks.

"We've got to get the plane to higher ground!" I struggled for breath.

The slippery gray-black shale worked against us. Frantically we moved the plane out of the ocean's grasp. I'd brought along gunny sacks for such a crisis and now we filled them with large rocks and used them for tie-downs. After one more round of testing the ropes, we grabbed our sleeping bags and made for the shelter.

Apparently getting caught in these storms was not uncommon. Someone had planned ahead and assembled a sod cabin. If mantled with snow, it would have looked just like an igloo. The round sides and top were made of sod. We had to stoop to get through the low door that leaned upwards, rather than truly vertical, but for all we cared, it was the doorway to heaven. Some Good Samaritan had provided us with cases of army rations, firewood, a Yukon (barrel) stove, and clean planking on which to throw down our sleeping bags. The thick sod-insulated walls shut out the howling storm and contained the easily lit heat. For the moment, we were safe and sound.

Emergency shelter (sod house) on Arctic beach, July 1959

By morning, the blizzard had worn itself out and visibility was up to one mile. Takeoff posed a problem. The high breakers covered most of the usable beach. To take off, I'd have to follow a tight line, with one wing over the water and the other over tundra. The waves shared less than 600 feet of the shore, forcing me to use the short field technique. I held the brakes, pushed the throttle full forward, and eased the stick back at only 45 mph. The rising shoreline bank forced me toward the water. Scarcely above stall speed, I lifted off the main gear, flying level above the cold waves — white-knuckled — until the plane gained a margin of altitude.

We remained at 100 feet and soon, through the light snow, I recognized the military installation of Cape Lisburne. We were flying so low that radar probably couldn't pick us up, but I was still uncertain about military clearance regulations, so I swung out over the water and around the rocky, windy cape, picking up a welcome 30 mph tailwind from the cloud-strewn sky. In this unpredictable north, I never knew what to expect: head winds, tail winds, military surveillance, blizzards, or sloppy beaches.

With the narrow protrusion of Point Hope moving into view, I was now in semi-familiar territory and expected to see recognizable faces when I landed. Sure enough.

"Hey, Doc, what are you doing up here?" Asked my hunting-buddy Leonard Lane.

It was terrific to see his face. He was a large teddy bear of a man who just made a person feel good to be around.

What a contrast from Barrow. The villagers genuinely inquired as to our trip through the bad weather, and then without a second-thought they filled our stomachs with food and the plane with gas. It would have been easy to spend more time in the comfort of their hospitality, but I needed to continue on schedule since I didn't know how long the flyable weather would last. After briefly stretching our legs, we were back in the air.

Between Cape Thompson and Kivilina, we were entertained by seal and beluga whale cavorting in the water, and long ivory-tusked walrus, which the storm had apparently ushered to the shore. The lower view added to our sightseeing and we were making great progress, nevertheless, I preferred more elevation and soared up to 4,000 feet. Kotzebue appeared in the distance. We'd flown over the other villages, but this time we'd land. I lowered full flaps and tried to settle on the middle of the runway.

"Doc, what are you trying to do?" puzzled Mel.

Every time I taxied more than a few miles per hour, the plane rose back into the air, like a kite. I couldn't figure out what I was doing wrong, but I sure was hopping down the airstrip. Then I remembered the FAA radio report of 25 to 35 mph winds on the runway. Upon releasing the flaps, the plane settled down into normal behavior.

"I told you planes naturally want to fly," I quipped.

We refueled and caught an FAA weather report.

"What's the good news?" ask Mel.

"Yesterday, a storm chased us, and today we're going on the heels of a storm. Hughes and Tanana have 1,400-foot ceilings."

"After what we've been through, that should be flyable!"

"Let's see how far we can get," I said.

The route I'd mapped out was untried by me, yet with potentially distinct milestones. We began our uncertain path back to Tanana by following the Kobuk River. Intermittent rain drizzled lightly on the windshield as we neared Kobuk. At this juncture we would turn southeast to find Hog River, which would lead us near Hughes. Unfortunately, we'd followed, and caught, the storm. The drizzle turned into a steady rain forcing us to scoot along at less than 500 feet. At this altitude, and with limited visibility in all directions, I was having trouble keeping my bearings. I'd taken Mel to be another set of eyes, and I really needed that right now. He didn't disappoint me.

"There's the Koyukuk River!" he pointed out. "Hughes is around the river bend."

Now we were both in known territory.

"Let's see if Mr. and Mrs. James are there. We can get coffee and a weather report," I said. I could use a break.

I didn't really drink coffee, except when that was the only hot beverage available, so I was thankful to see hot cocoa. I sipped the steaming beverage and paced around their cabin trying to figure out what to do. We were so close to home, only 90 miles away, yet two pilots had tried to get through shortly before we arrived and were forced back. Conversely, I thought I could make it through. I'd flown this area many times and I was counting on the 800-foot pass to be open with a 1,200- to 1,400-foot ceiling. Besides, I rationalized we could always turn back if we needed to.

"Mel, let's refuel and take it on in."

We skirted the south edge of Indian Mountain and the plane moved among the eastward hills. All we needed was one pass that would take us to the west headwaters of the Tozi River. We picked our way into the passes. Every time we had to turn around. Time and time again we fumbled around as thick clouds met us in canyon after canyon. The low ceiling added to the confusion and blurred any distinguishing ridge-lines. I finally gave in to my frustration.

"I'm afraid we're going to have to head back to Hughes."

Mel had no comment.

I turned the plane around and worked my way back. To my alarm, we were entangled in the labyrinth of green hills and shallow canyons. In and out and around we flew for nearly an hour. Everything looked the same.

"Let's face it, Mel," I confessed with exasperation. "We're lost, and there's no place to sit down."

Black-green spruce, thick birch, and lofty cottonwoods all poked up at us.

"Yep, it's a mighty tight spot, Doc."

My overload of anxiety was interfering with my thinking and I knew I was on the verge of sheer panic.

"Mel, we could really use the help of the Almighty Copilot."

I prayed with my eyes wide open and Mel followed with another heaven-directed SOS. For the moment, this had a calming effect, and together Mel and I analyzed our plight. Then, I realized that our plane compass showed 330 degrees, and it should have read 90 degrees. Something was wrong. My mind started to freeze again, but then I realized my foolishness. All this time I'd been relying on only my own sense of direction when I could have used my radio. I tuned into Bettles with the Bendix loop, an old World War II navigational direction finder, and learned that the compass *was* correct, but we were heading north to Bettles, rather than south to Tanana.

"Now that we have our directions straight, I'm going to follow the first stream I find to get us out of this maze."

Mel didn't seem to be listening, but stared intently past me and out my window.

"Hey, isn't that the burn area we saw the first day of our trip?"

Sure enough. I recognized the blackened hillside. "We must be on the northern slope of the Ray Mountains," I choked out a laugh.

The good news kept coming. While we'd been searching the ground for landmarks and direction, the ceiling had lifted in spots to 2,500 feet. There, just in front of us, appeared to be a shallow pass, with a slight 1,000-foot clearance. I headed toward it and squeezed through the gap. We were chalking off the obstructions. Our remaining problem was to get through the 4,500- to 5,000-foot Ray Mountains where ragged white clouds covered the peaks.

The Bettles' radio helped me decide what to do.

"Tanana has a 2,500-foot ceiling, wind 20 knots from the south."

From this information, I knew the south wind was pushing and bunching the clouds against the south side of the Ray Mountains. I had read that at times, beneath these conditions, the wind would push a hole up through the clouds.

"I'm going to try to find an opening along the north side of the peaks," I told Mel.

Suddenly, through a small gap, I caught a glimpse of the valley on the other side of the mountain.

"Here we go," I yelled.

I cut the throttle and began slipping the plane down the hole to the valley. The plane gained speed and rapidly lost altitude. Within minutes, we burst out under the 2,500-foot cloud layer.

"We made it!" I shouted in unrestrained relief.

The altitude was sufficient, but we'd need more than that. Visibility would have been nice. Rain came down in torrents. We both strained our eyes downward. Amazingly, without much effort we identified the Tozi River and managed to call into the Tanana FAA.

"Mel, good weather conditions ahead!"

The adrenalin was surging for a different reason now. In three days and five hours we'd covered 1,300 miles.

In a letter to my parents, Ruby summed up the experience:

> *The highlight of the week was Elmer and Rev. Jensen making a trip from here to Barrow, Point Hope, Kotzebue and returning Saturday night. They had contrary weather the greater part of the trip, which made for reality mixed with a nightmare. I will not go into detail as I want Elmer to write it up and then it will be very interesting for you to read.*

This unpredictable and rugged land showed no favors to the weak or fainthearted. I loved Alaska, and in moments like this, after conquering its grasp, I wondered why I was leaving its challenge and beauty.

CHAPTER 21

OUTSIDE

August 1959

RUBY AND I HAD VENTURED to the Alaska Territory as wide-eyed Kansas farm kids, unaware of the realities. Now, cognizant of the no-second-chance hardships and dangers, the allure had only increased; yet we were heading away from what felt like home and had subsequently become the State of Alaska.

As we'd approached Tanana, on our way in from the Barrow expedition, Mel Jensen had questioned me about going Outside.

"Yeah, I'll really have to pack away the hunting and flying gear now," I said. "Not much time left.

I'd flown low, circling the village. My eyes and mind tried to collect these last images of the Yukon River and the familiar Athabascan village. Stumpy black spruce, draped with untidy charcoal moss, preceded the village. Then the FAA housing came into view, followed by the medical facilities. A second later we were over the cabins with busy washing machines on front porches and smoke half-heartedly drifting out of stove pipes. Children played on Back Street. Huskies, their tails curled, stood on top of their houses. I swung out farther than usual. The gravel pit, with mountains of aggregate, brought to memory the pleasure of my children on infrequent walks to that outside edge of the village. They'd climbed, slid, and shrieked in amusement. Then there was the cemetery with the elegant shake-shingle roofed Mission Church, its belfry reaching to the sky with a small white cross on top. Tall grass and fireweed obscured the cemetery fences, while stately cottonwoods surrounded the church. I banked into a turn and started my descent. A fish wheel with long poles clutched the river's bank, and swept up unsuspecting fish.

After the terrain I'd landed on the past three days, this airstrip was perceptibly agreeable and wide.

Mel had unlatched the Cruiser's door and I climbed out after him. We stretched. It had been a harrowing day. Overhead, seagulls squawked and black crows cawed raucously. I breathed deeply the ever-present smell of wood-stove smoke which wafted around the village.

The two years in Tanana had its good and bad, difficult and rewarding. In a letter to Public Health Service supervisors, written later, my wishes were evident:

My initial contact with the PHS was very satisfactory. My two years in Anchorage were profitable. I appreciated the acceptance of my request to transfer to the field hospital in Tanana...

I missed contact with fellow colleagues and the opportunity for refresher training...

I was disappointed when I was asked to transfer to the main States. There was no consideration given my family. I asked to continue one more year, even though the typical rotation is two years. With this denied, I requested a transfer to Cass Lake, Minnesota, where it would be possible to obtain post-graduate study.

Of various assets derived from my work at Tanana, I learned to supervise, respect others, and appreciate teamwork. The living quarters and medical facility were satisfactory, as was the work load.

Unsolicited by me, my transfer was to Browning, Montana, a Blackfeet Indian reservation near the border of Canada.

I wasn't alone in my frustration. Ruby had no desire to leave either. She'd more than survived in this isolation that had few daily living amenities or opportunities for social stimulation. She and Anna Bortel had resourcefully, humorously, and determinedly made the most of village life. Ruby had put together comical, as well as fancy, social events, helped organize Tanana's first library, and initiated picnics — in summer *and* winter. She'd embraced the Native women and tried to discover from them how to live in the wilderness. She'd endeavored to learn their bead-work; and she'd acquired a taste for smoked salmon strips.

Added to her weariness, Ruby had made home in three different houses — in four years. Then, as if packing wasn't arduous enough, there were terms for how to pack. Our belongings would be transported by barge to Fairbanks and then by truck to Browning. The stipulations required everything be in containers suitable for one person to carry. Heaped on top of that pressure, the guarantee of things arriving safely in Fairbanks was questionable. If inclement weather occurred, the decks on hand would do nothing to protect our shipment.

The entire family balked. Naomi and Ruth cried, reopened boxes, and pulled out their dolls, little stove and dishes, and found the round cans of letter cubes to play Spill & Spell. All this resistance would have made a less durable woman throw her hands in the air, crumple in a heap, and bawl.

Just as I'd flown the J-3 from Anchorage to Tanana, my piece in the moving process was to fly the Family Cruiser to Browning — down the Alcan Highway. I thought it wise to have a flying companion, but that required someone willing to fly with me, and to find transportation back home by some other means.

On August 2, 1959, I wrote my parents:

Everything is crated and our cupboards are bare. Last night we finished marking and weighing the crates. We had about 4,650 lbs.

The Lord has answered our prayers for a plane passenger for me. He is a young married man we knew in Anchorage. I'll meet him in Nenana Monday morning next week as I'll be seeing the family get on the train at Nenana. They will go to Talkeetna for a few days to visit with the missionary Scripters and they'll go to Anchorage for about five days visit with friends.

The new M.O.C. arrived Friday. He is from Kentucky and just finished his internship. His wife and 2 yr. boy will come in a couple weeks. He of course is a commissioned officer but expects to only stay one year. He appears to be fairly level-headed. I've turned about everything over to him already so I can enjoy a week of "vacation." I'll get in some fishing, and help work on the old school and then I'll be ready for that flight down the highway.

There were many loose ends to tie up; in lurching steps forward and backwards, we proceeded. One detail was property. Anna had purchased a village lot, and so had we. Ours was of irregular dimensions with a mix of wild grass, scrubby bushes, a wild rose bush or two, a few short aspens, and a drainage ditch running across it. Nevertheless, there was something consoling about having that piece of Tanana, like keeping a foot in the door for a homecoming — even though we had no idea if or when we'd return to Alaska. Regardless, we felt no urgency to part with this parcel.

On August 7, 1959, I wrote my last letter from Tanana to my sister, Lillian:

Our house is stripped except for a few furnishings. We're practically living from tin-cans now. We have quite a few picnics. Today three of us families went blueberry picking about ten miles from here. We only got a couple gallons.

Last night I took Rev. Jensen, Wally and his boy, to Minchumina (Min-CHOOM-in-a) Lake, 95 miles from here. There we fished for pike in the same place I took folks last summer. Mt. McKinley was perfectly clear shining in the evening sun.

I'm getting the plane in good condition for the long run. I've worked out the flight plans nearly to completion so it'll be an enjoyable trip without much sweat.

The weather is finally clearing and now we have normal summer weather with daytime hi of 64° and nite low of about 40°.

Within hours apart, I took off in the PA-14 to Nenana, and Ruby, Naomi, Ruth, Mark, and Mishal flew by charter. Again, we ventured into a place of unknowns; hoping for the best, yet wary of subtle forewarnings.

Browning, Montana

The Gaede's relocation to Browning, Montana is written about in "From Kansas Wheat Fields to Alaska Tundra: A Mennonite Family Finds Home"

Tanana Lot

Lots Six and Fifteen, Block Six, Townsite of Tanana, records of the Ft. Gibbon Recording District.

Parcel I containing 52,673 square feet more or less.

Parcel II containing 2.08 acres more or less.

Lot Six was deeded to the Yukon Telephone Company in 1970.

Lot Fifteen was sold to Paul and Mary Starr in July 1997.

A picturesque cabin sits on this lot, which is adjacent Paul and Mary Starr's cabin. Their daughter enjoys her time there as often as possible.

CHAPTER 22

BACK HOME

July 1961 to 1966

September, 1962

Wilbur Hett (Friend from Hillsboro H.S. Glee Club)
Hillsboro, Kansas

Dear Friends,

I'm spending the night at our medical clinic, keeping an eye on a set of three-pound twins born at midnight, so I thought I'd give you a rundown of what I've been doing.

WE'D JUST RETURNED TO ALASKA.

Nine months at the Public Health hospital on the Blackfeet Indian Reservation in Browning, Montana, was all I could take. The patient load, even with three of us physicians, was crushingly unrealistic. It wasn't only that. I'd been slugged in the face by an inebriated patient. Naomi, distressed by her school situation, went into childhood depression and nearly stopped eating. Ruth, on the other hand, coped well and, in an unexpected display of confidence, had volunteered to play accordion in front of her third-grade class. Mark and Mishal were unfazed. Mark had snow — and he was happy. The summer and winter wind wore on Ruby's nerves. She could scarcely keep clothes hung on the outdoor line. Unlike Tanana where snow drifted down softly, here, the unremitting wind packed it hard enough to carve igloo blocks and dig tunnels, suitable for Mark's play, but maddening for Ruby. In winter, when she tried to dig our new and modern VW microbus out of the detached garage, every shovelful blew back into place — or her face. Local kids climbed on my airplane, which didn't help my attitude, not that I had time to fly it anymore.

I resigned from the Public Health Service and sought other opportunities; opportunities for refresher training and post-graduate work. One such place was the County Hospital at Tulare, California, where I could do a general practice medical residency. The clincher had been that my parents, brother, sister, and their families lived a short distance away, at Reedley and Fresno. I couldn't pass up this opportunity for our family to integrate with extended family. Mark and Mishal didn't know much about relatives, whereas Naomi and Ruth carried affectionate memories of their grandparents with them. And, in this milieu, the entire family flourished with the continuous get-togethers with cousins, aunts and uncles, and grandparents. We couldn't get enough of drippingly sweet watermelon, tooth-chilling homemade ice cream, picnics — without mosquitoes — swimming, trips to see giant redwoods, and the wonder of Disneyland. We lived in Fantasy Land.

Even with all this, I was restless. In fact, Ruby and I both wanted simpler living, service to others, and, in each our own ways, more of a challenge.

Among other possibilities, I spotted a medical journal advertisement placed by Dr. Paul Isaak, sole physician in the Soldotna (Sol-DOT-nah) area, on the Kenai Peninsula, requesting help with his growing practice. I inquired and waited for an answer. If he accepted, he and I would be the only full-time physicians in the area. The nearest hospital was 98 miles away by road and 65 by air.

The entire family knew I'd sent the inquiry and now waited a reply. It arrived in the mailbox on the broad and blocky front porch of our bungalow. I read the letter at the supper table after work. Then looking around the table at my family, Naomi (eleven), Ruth (ten), Mark (five), Mishal (three), I cleared my throat.

"Looks like we're heading back to Alaska!"

We'd be saying farewell to family, friends, the fruit-fragrant climate of the San Joaquin farming valley, and a Dick-Jane-Sally neighborhood with sidewalks and paved streets, tidy yards with flower beds, a grade school two blocks away, a nearby library, and grocery stores overflowing with milk, bread, and fresh produce.

As in previous relocations, my airplane complicated the transition. I still had the PA-14 Family Cruiser, which I would now be flying up the Alcan Highway. Flying had been pretty tame in the Lower 48 — less adventure for me and more entertainment for guests — and I looked forward to literally stretching my wings again. One relative, Wally Loewen, volunteered to fly up with me. Given he was a glider pilot, he obviously liked living on the edge, and to him, such a trip sounded more exhilarating than fearsome. I flew ahead of the family, but returned commercially to drive up with Ruby and the children. My cousin got his money's worth of thrills on the flight up.

Ruby and I had driven this route in 1955; now six years later our VW microbus toiled more noisily and grudgingly than our 1947 Chevy had. Nothing much had changed in regard to the road, and there were plenty of dust and pot-holes left over to make this trip live up to the reputation of driving the Alcan, usually said with a shake of the head, heavy sigh, or wincing of the face, all of which preceded some narrative designed to beat the other fellow's tale of traveling woe.

This time, we had double the children, along with a tiny, often annoying, Pekingese dog. This creature had been acquired without my recommendation when Ruby's parents showed up on our doorstep in Browning, and the children rushed to hold it. Mishal had shrieked in unrestrained delight and nearly pulled the black pup into pieces. I was without words. All in all, instead of four of us driving up the abusive Alcan, there were two adults, four quarrelsome children under the age of eleven, and a yipping dog. Fortunately, none of these were finicky or fussy about sleeping accommodations or meals, of which all were met by the VW, a tent, or a picnic table.

I had become acquainted with the Kenai Peninsula when I'd taken floatplane lessons in Anchorage. During my practice time, I'd flown over the Swanson River oil discovery area. I'd heard there was enough gas in the nearby wells to supply all of Alaska. At this time, 1961, oil wells produced 30,000 barrels per day and an oil refinery was to be built in spring, north of Kenai.

The need for another physician was the oilfield workers and their families. Families required general medical care and workers dealt with the notorious oilfield reputation for accidents. These emergencies would increase in 1962 when offshore oil was discovered in Cook Inlet and workers flew on and off the platforms.

Whereas previously my adventuresome spirit had sought out smaller, inaccessible vicinities, my focus had shifted to economic feasibility. I'd be on my own now, not under the umbrella of Public Health. The Kenai Peninsula was projected to have the most rapid growth with the best future development potential in Alaska. The hub of activities seemed to be Soldotna, with around 332 people living in log cabins on their homesteads, or in trailers and rough-sided shanties in what was considered to be town. Soldotna was at a crossroad with one gravel road, the Sterling Highway, continuing to Homer, and the other, the Kenai Spur Highway, to Kenai, 12 miles away. Kenai was comprised of 778 people and had an area laid out like an actual settlement. The population of both areas was a surmised count since many homesteads were far from roads and tucked into pockets of cleared land or hidden within the forest.

Besides oil production, the economy was based on commercial salmon fishing. Wilson's grocery store extended credit for people who could only pay for groceries after fishing season.

Clinic-House, 1960

Prior to my arrival, Dr. Isaak had redesigned one of the few frame houses in town to be his clinic. It was located conveniently on the Sterling Highway. He'd used the closet-size bathroom for a laboratory and had an X-ray machine in one of the two bedrooms, of which the other was an examining room. He, his wife Amy, and their five children lived in a trailer beside this make-shift medical facility. A rudimentary airstrip ran behind the house, making it convenient for Dr. Isaak to park his Stinson airplane and fly it to the hospital at Seward when necessary. (In the future, the airstrip would become Wilson Street since it was also behind Wilson's grocery store.)

Prior to my arrival, several local businessmen had built a two-story, concrete block medical clinic along the Kenai Spur Highway. About the same time, the Isaaks moved to their homestead, freeing up the green-trimmed, partially sided house-clinic. Accommodations in town were scarce and we gratefully rented the small house, even though it had little or no insulation, and any heat generated by our oil furnace escaped through the walls.

The cross-roads town was booming. A new grade school had just been built; even before its completion, the projected attendance had spiked. In fall, Ruth's class was relocated to space in a low, flat office building.

Central Peninsula Clinic, Soldotna 1961

The clinic was fairly complete and had an excellent X-ray unit and laboratory, along with rooms for minor surgery and obstetrical deliveries. Minor surgeries included tonsillectomies, adenoidectomies, and D and Cs.

Understandably, it was tricky to focus on performing surgery and monitoring anesthesia all at once, and Paul Isaak welcomed my adeptness with anesthesia. We were a competent team. Between surgeries, emergencies, and minor ailments, we delivered babies at an astonishing rate — the population was exploding beyond that of mere relocation.

There was a ramp on the side of the building for stretcher cases, some of which were flown in by helicopter from the oilfields; subsequently the gravel parking lot served both land and air traffic. The walk-out basement accommodated a dentist's office and a new library. Just like Ruth's class, Naomi's grade school class flowed outside the school perimeters and was shoe-horned into a room adjoining the library. This arrangement suited Naomi, who was an ardent reader of historical biographies, which lined the lower shelf just outside her classroom door.

In addition to our local medical work, Doc Isaak and I held weekly medical clinic at Seldovia (Sell-DOE- vee-uh), about 110 miles south and accessible by airplane or boat only.

When I wasn't going to find patients, they found me — and not always at the clinic. It wasn't unusual to hear patients walk up the short boardwalk in front of our house and knock on the door after hours. This wasn't only because they were used to the house being the clinic and a doctor available for medical needs, but telephone communication was non-existent to many. Once they'd driven the miles into town, it was just as easy to make a direct stop. Besides, it wasn't as though I had an unlisted number. All in all, I was easier to access than my partner. To find Dr. Isaak meant a mile drive into the wilderness, oftentimes maneuvering on nothing more than a mud-bogged trail or ice-bound track.

After seven months, my family and I moved to a newly built two-story house on Corral Street within sight of the clinic. This served me well. I could run home for supper yet watch for an emergency to arrive by car or by helicopter.

In Montana and California, I'd used the VW bus more than the Family Cruiser. Now, back in Alaska, the airplane resumed its status as a fundamental form of transportation — rather than merely a source of pleasure. With this reliance on aviation, my life resembled Tanana. In a letter to a Kansas friend this was evident:

Saturday noon, Ruby and I flew over to Nondalton, 180 miles southwest across the Cook Inlet and through the Alaskan Range. We visited with the Arctic Missions missionaries and the schoolteachers, and did physical examinations on all the school children.

Monday, we flew home and I was back in the clinic by 3:00 p.m. My second patient presented himself with abdominal pain. After complete evaluation, Dr. Isaak and I decided he must have immediate surgery for acute appendicitis. The patient was taken by private car to Seward General Hospital and since it was dark and the weather was not suitable for flying, Dr. Isaak and I also drove the 98 icy miles.

Tuesday, we treated 46 patients.

Wednesday, at 8:00 a.m., we took Dr. Isaak's plane and flew to Seward, where we did two surgeries. We returned at 3:00 p.m. and worked in our clinic until 6:30 p.m.

Thursday was a routine clinic day and we had one obstetrical delivery.

Friday was clinic day at Seldovia. It was my turn and I left at 8:30 a.m. I flew across the wooded country and over 18 miles of water to the fishing village of about 400 people. Most of the people are Athabaskan

Indians, who earn their living from the local salmon and king crab industries. I offered rides home to two live king crabs. They were over three feet in diameter and crawling all around in my back seat like giant spiders. I stuck them in the bathtub when I got home for the children to see. Ruby experimented with open-faced toasted crab sandwiches. I finished clinic at 6:00 p.m. and flew back in the bright moonlight. I could see nearly 70 miles.

Saturday morning, I flew Naomi to the nearest orthodontist, at Palmer, which is about 110 air miles away and on the other side of Anchorage. After buzzing the dental clinic, the doctor's wife came to the airfield and took us to their clinic. By noon, we were back in Soldotna.

An aspect I hadn't mentioned was that I'd been roped into veterinary medicine. One routine day at the clinic, my petite and lively red-haired receptionist caught me in the hall between patients. Her face was a mix of amusement and bafflement.

"Dr. Gaede," she said in a hushed tone. "You have a patient tied to the guardrail at the end of the emergency ramp."

Her brown eyes twinkled and she laughed at the perplexed look on my face.

I opened the door. There at the bottom of the ramp was a horse tied to the rail and a young woman in her late 20s. She spoke when she saw me.

"Dr. Gaede, there isn't a veterinarian in this area. I thought maybe you could look at my horse," she pulled her wool jacket closer. Her hopeful eyes peered beneath curly bangs.

I could only trust that my farming background would augment my knowledge of medicine.

The horse's leg was the presenting problem and I managed to examine it without getting kicked. My diagnosis was an infected barbed wire cut, for which I prescribed antibiotics and an ointment. At that point veterinarian medicine seemed simple enough.

Another time, I walked into the waiting room and discovered that my next patient was a spider monkey. The people-patients didn't seem to mind the diversion, and I suspect it temporarily alleviated their waiting-for-the-doctor anxiety. The monkey's master, a woman with three small children — all crying in various degrees of hysteria — explained desperately that the inquisitive creature had swallowed a fishhook. Although gagging off and on, their pet seemed unperturbed by the obstruction and leaped about on patients, end tables, and even the receptionist's tall check-in counter.

I was well-acquainted with fishhook cases. Over the summer both experienced and novice fisher men and women accidentally became fishers of people, rather than fishers of fish. This, however, was the first case when the hook was in the throat — of an animal.

I firmly grasped the furry nuisance and took the wailing entourage into the operating room.

"Meadows, bring me a pillowcase," I told my nurse. Her first name was "Jeanie," but somehow, I'd gotten into the habit of referring to her — and my partner, Paul — by their last names. In front of patients, I usually add "Miss" or "Dr," but at this moment, in the unusual situation, I forgot myself. She turned on-heel and returned within the minute. She wasn't new to innovative medical procedures in the north.

I stuffed the flailing monkey in the pillowcase, tied shut the open end, and began dripping ether on the cloth over the monkey's head.

At first, the children stood spellbound, but then resumed their bawling. "Monkey die! Monkey die!"

"The monkey will be fine," I assured them. "He will fall asleep so I can remove the hook from his throat."

Just as I said, my patient stopped thrashing. I removed the pillowcase from his inert body, pulled out a long forceps, and removed the hook.

Almost immediately, the monkey awakened from his short nap and jumped around the room, chased by the children. I'd been known to tell Ruby that my day at the clinic had been like a zoo, but today I was really telling the truth!

These animal stories were fun to entertain guests around the Sunday dinner table, but the essence of life on the Peninsula entailed hard work. We were one of the last families to get in on the Homestead Act. We "proved up" a flat, heavily wooded 80 acres on Gas Well Road, about three miles west of the Soldotna bridge.

The first time we walked about this parcel, a bearded old-timer, flashing a large pistol on his belt, stalked out of the woods. The gruff man wore rugged army clothes and was over six feet tall.

"Who are you and what are you doing here?" he barked.

I answered his questions, but my reply did not seem to please him. He shook his fist and muttered something about us just trying to get this land. I got the feeling he had someone else in mind for its possession. We found out later that Woody lived a half-mile south of us, and although he was threatening at our first encounter, we had no trouble after that. I believe he liked to maintain his desperado image.

The land had no lake, no view, not hill, no riverfront, but it was ours. I thought back to when I was a boy. We'd lived on nine different farms. My parents always hoped and tried to better themselves so they could afford their own land, but it was pretty much an impossible dream since the landlord took one-third of their crops. They could never save enough until I was in college. Regardless, as a farm kid, I'd grown up with the land under my feet, and dirt between my toes and under my fingernails. I understood the kinship between a farmer and the earth. Now I was eager to make this piece of wilderness my own.

We were given three years to "prove up" the homestead, which specified clearing one-eighth of the land and planting an annual crop. Sixty-foot spruce stretched toward the sky with shiny green-leaved cranberry bushes gathered at their feet. Old windfall left rotting stumps, decorated with curling light green moss and soft gray fungus ledges. Golden aspen and cottonwoods filled in the space between.

Unlike some homesteaders, we were fortunate to have immediate access to electricity. Another bonus was the water table, only 12 feet below the surface. We easily hand dug through the sandy soil to drive the well points. Gas Well Road was a solid gravel road, which could seem insignificant to a newcomer; however, a gravel road versus a dirt road made the difference between walking or driving a mile or two during spring breakup.

Ruby and I cleared 14 acres by chainsaw and axe, cutting wedges in these immense trees, then slowly pushing with all our might — or hanging off a lower limb to pull them off balance, until they crashed down into the quiet forest. The laborious process moved slowly over the days and nights of two and a half years. After cutting down several trees, we would trim off the branches and start an enormous bonfire to clean up the debris. We primarily worked in the winter since wildfire was always a fear.

Cabin on the Gaede Eighty Homestead

Although our main house was a bi-level with brown-stained siding, we peeled some of the straightest trees to build a shiny, varnished log cabin and red tin-roofed horse barn, complete with hay loft. To further enable our homestead construction, we used a portable sawmill and cut between 30,000 and 40,000 board feet of rough-hewn lumber, which we used to build a hangar and woodsheds. We worked on my day off and Saturdays, often by moonlight and the light of the bonfires.

At times we managed to capture the children's energy and direct it to peeling the logs, but usually they played away their natural energy resources by following the leader and walking heel-toe around the wonderful pathway maze of long, fallen logs. From the air, our clearing looked much like a giant game of pickup sticks.

After our hands-on work, a D8 Cat pushed out the stumps and leveled the land. The stumps were shoved into piles, along with the rich black topsoil that clung to their roots. These stump piles could be seen along all the Peninsula back roads and cleared areas. A number of people owned horses and tethered their horses to the piles where the rich soil, produced lush green grass. Bright pink fireweed and purple-blue lupines thrived amongst the tall grassy stalks.

This work was not only to meet requirements, but for a specific purpose. When all brush and timber was cleared away, we had a half-mile-long by 120-wide foot long runway a half block behind our house. Our Herculean accomplishment showed up on aviation charts as the "Gaede Private."

Gaede Private airstrip

In further compliance with the homestead requirements, we planted annual crops of oats and then timothy hay on this airstrip. It proved true, "You can take the boy off the farm, but you can't take the farm out of the boy." My farming instincts took over and before I knew it, I'd purchased a small Ford tractor for harvesting.

Harvesting was a family project of cutting the hay, then raking it up and pitching it on a wagon. Although it would seem we had a large field, Molasses, our brownish-black mare, could never have survived the winter on such scanty rations. We ended up supplementing our hayloft with other folk's hay.

Our pet population seemed to explode on the homestead. We collected a menagerie of pet rabbits, orphaned lambs, and always some large dog and a horse — in addition to the tiny dog we'd brought with us. Chickens appealed to Ruby, and then goats, geese, and ducks. She was the ultimate farm girl. As far as we were from neighbors, our cats managed to court trouble and, in an attempt, to control this prolific population, I neutered this multiplying species. Ruby assisted me. The surgeries required anesthesia. Cats differed from people, however, and I had more difficulty calculating the amount of medication needed. Nothing was more exasperating than to be halfway through surgery and have a cat leap off the operating table.

There were other surprise outcomes as well; in particular the situation with Yofee, Ruth's cat. I was determined this cat would not interrupt my surgical endeavors. Ruby and I managed to hold down the calico cat, while I administered the anesthesia. Surgery went well and Yofee remained asleep.

"Okay, Ruby, I'll undo the IVs and let's go home," I said, removing the needle.

The cat didn't stir. Much to my dismay, I realized that the intravenous needle had become dislodged and some of the anesthesia had escaped into the muscle tissue. I'd overdosed the cat. Carrying the motionless, yet still breathing, cat in an apple box to the car, I brushed off Ruby's questions with, "I do the surgery. You deal with recovery."

As soon as we got home, Ruby placed the cat and the box on the floor by the low picture window and assessed the situation. The cat's eyes were semi-open. She pressed them close, but they returned to a fixed stare. She resorted to covering them with a moist cloth. Her next concern was dehydration. She threaded a tube down Yofee's throat and poured in eggnog. Her intensive vigil continued off and on through the night, and the next night, and the next night.

I remained unconcerned. If anyone could revive the cat, Ruby could. After all, she was the only "nurse" I could get to come out for middle of the night deliveries at the clinic; and she capably assisted me with roadside car accidents. At other times, she kept out-of-town missionaries or school teachers in our home for post-operation or post-delivery care.

One week later, Ruby and I were enjoying an early morning breakfast of sourdough pancakes and moose sausage. Yofee passively joined us in her convalescent box near the end of the table.

"Elmer, the cat is doing something," exclaimed Ruby, jumping up.

I looked over. The cat was exhibiting the initial reflex response of swallowing — and about to swallow the food tube.

"Yofee is coming back!" Ruby said jubilantly.

This marked the beginning of Yofee's second of nine lives.

Ruth wondered how Yofee would act after her lengthy voyage into the unknown. Truthfully, Yofee was not one of our more intelligent cats and we noticed no difference in her behavior.

Homesteading was a new chapter in my short Alaskan history, but the challenge of mixing medicine with flying remained much the same as it had ever since I'd arrived in Alaska. During some weeks, Dr. Isaak and I would make several trips to the Seward hospital in his Piper Pacer or 180 Cessna, or my Family Cruiser.

Seward, a year-round seaport town on Resurrection Bay, curved around many docks and was nearly pushed into the bay by the mountains. "Resurrection" seemed a strange name for the bay, the river flowing into the bay, and the pass through the mountains into the town, but once I had heard the story of all their namings, it made more sense. Apparently, Alexander Baranov, a Russian fur trader and explorer, had met rough seas on his voyage between Kodiak to Yakutat. Unexpectedly, he discovered shelter in this bay and because it was the Russian Sunday of the Resurrection, he named nearly everything resurrection.

In spite of Baranov's influence and profuse and redundant names, Seward bore the name of the man who was instrumental in purchasing Alaska from Russia in 1867, U.S. Secretary of State William H. Seward. Fortunately, this frontier town didn't carry the original nickname of the Alaskan purchase, "Seward's Folly" or "Seward's Ice Box."

We pilots had names other than "Resurrection" for the pass that capriciously allowed access to Seward. The valley was a cauldron for nasty weather and fog. Added to these unwholesome conditions, a tremendous wind often swept along the Anchorage Highway into Seward, which nearly blew cars off the road below and created severe turbulence above. Subsequently, transportation to the hospital in life-or-death medical emergencies, was often complicated and nail-biting.

Such was the case one particular late summer day. Dr. Isaak and I had been busy delivering babies between checking ears and throats of other patients, casting fractured legs and arms, prescribing antibiotics for various infections, and tranquilizing dogs to extract porcupine quills. As if in a relay race, I moved between patients. Before entering another closed-door examining room, I grabbed the next medical chart: Sherron Justis, 12 years old.

Sherron's mother filled me in with the details. The girl hadn't felt well the previous night and had developed a low-grade fever, nausea, and vomiting. She also had right lower abdominal pain. After examining her and doing some blood work in the lab, my suspicions were confirmed. I decided she had acute appendicitis and would need immediate surgery. I tapped on the examining room door where Dr. Isaak was with a patient, listening, diagnosing, and treating as intensely as was I. We conferred for a moment, and then asked the receptionist at the front desk to call the Seward hospital to expect our arrival in 90 minutes. Her next task was to reschedule the rest of our day's appointments — as if there were many spots for them to overflow in to the following day. Some of these people had driven for miles, or even flown in, so she'd be facing a disgruntled group.

Sherron, her mother, Dr. Isaak, and I slipped out the back door and slid into Dr. Isaak's Mercedes. Instead of taking Kenai Spur Highway, the main highway through town, we short-cut through the back streets, crossing over the old airstrip, and catching the Sterling Highway in front of the house-turned-clinic. My plane was tied down at the recently completed Soldotna Airfield, which lay across the Kenai River, several miles out of town on Funny River Road.

We never knew what waited for us between here and Seward, but at this moment the weather looked pleasant with high furrows of clouds in the calm skies. Mother and daughter found their spots in the back seat, and Dr. Isaak, who had logged pages of flying hours, sat beside me as co-pilot. I confidently went through the takeoff checklist, opening the throttle and checking the magnetos. Everything sounded fine. Within moments we were airborne.

I never tired of the scenery on trips to and from Seward. First, we flew over Skilak (SKEE-lack) Lake, a nearly 16 mile-long, glacier-fed milky turquoise lake. As pilots and hunters, we always looked for a place to set down in case of emergency — and wildlife.

"There's a nice one, Elmer," said Paul Isaak, pointing out the window to a moose in the marshy shoreline.

We climbed upward into the mountains and neared Upper Russian Lake, now scanning for sheep and goat in the crags, and black bear eating blueberries on the mountainsides. I glanced back at my passenger and patient. The girl slumped with eyes closed toward her mother, unaware of the prospective wildlife tour outside her window. I'd expected more complaints with the seriousness of her condition, and I was glad she could find some comfort.

Just past the lake, I noticed my engine RPM was dropping slowly. I suspected carburetor icing and casually applied carburetor heat. No change. The needle gradually slid around from 2,300 to 2,100 to 2,000, indicating that the engine was slowing down and that the propeller was making fewer rotations. As the mountains ascended, I descended to 800 feet above the thick spruce trees. During the past several years, several planes had crashed in this area because there were no emergency landing areas.

I turned to my veteran copilot. "Paul, I'm losing engine power. I'm sure it's carburetor icing, but my carb heat isn't doing any good. Any suggestions?"

He shook his head. The trees reached toward us with only a 500-foot clearance. If we crashed, we'd all have a medical emergency — and not enough doctors to go around.

Out of the blue, I thought of something I'd read about carburetor icing: when the carburetor heat doesn't remove the ice, lean out the fuel mixture. The engine will run rough since it won't have a balance of oxygen and fuel. Doing this for a few seconds will shake out the ice. My hands trembled as I reached for the smooth red knob. I drew back. I'd lose even more RPM if I leaned the mixture.

I looked outside below. I couldn't believe that Alaska had such giant trees — what were they doing so close to us?

We had nothing to lose. I grasped the knob. The plane shuttered. "Come on," I croaked hoarsely.

Seconds passed. Abruptly the engine sounded different. With a roar of full power, we defied gravity and climbed until 5,000 feet bridged the gap between us and the greenery below. My heart skipped a beat, but didn't slow down. I felt relieved, but not enough to casually resume looking for wild game; instead, I focused on the gauges, the sound of the engine, and any remotely possible emergency landing spots. After a while, I adjusted the mixture to its previous setting and pushed in the carburetor heat.

About ten minutes later, the nightmare started all over with the slowed RPM. I repeated what I'd done earlier. Even though I knew what worked to keep us in the air, it was a nerve-wracking experience. I was definitely getting more than the daily recommended dosage of excitement.

The trip seemed longer than usual, but eventually we landed in Seward and eagerly thrust our shaky-kneed legs out of the plane and onto the solid ground. A car from the hospital met us — as did the mechanic who had worked on the plane a week before.

The man stepped out of his truck. A wrench poked out of one pocket in his striped coveralls, and an oily rag from another. I looked him square in the eye.

"Ollie, we almost left this plane in the canyon."

He stared at me in horror.

"We have to go to the hospital for an appendectomy, but I want you to take this plane up and check it out before we get back."

I explained what had happened. The blond man with grease smudges on his face stood aghast.

Surgery went well, and we removed a hot appendix, which probably would have ruptured if we had waited much longer.

Two hours later, Dr. Isaak and I returned to the airfield. Our two passengers would remain in Seward. Ollie stood by the plane, but would not meet my eyes.

"What did you find?"

I had to strain to hear his reply, which was mumbled in a trembling and apologetic voice.

"I crossed over the hot and the cold hoses. You weren't getting any heat to your carburetor to clear the carburetor ice."

I didn't know what to say and just turned around.

Paul and I pulled ourselves into the crimson aircraft without a word. Within minutes the plane soared into the air and climbed out over

Resurrection Bay, before turning and swiftly making its way up through Resurrection Pass. I didn't care how much assurance I should have felt, I took the plane to 5,000 feet and kept her there.

After a half-hour of silence, I glanced at my co-pilot. "Isaak, this trip seems to have given you a few more gray hairs."

Not to be outdone, he replied. "Hey, I noticed you look a little gray around the edges, too, Gaede."

I'd earned them all, and probably had a few more coming.

CHAPTER 23

THE DAY THE EARTH BROKE APART

March 1964

THERE WERE NO PRESSING medical needs on this Good Friday holiday, so Dr. Isaak and I decided not to hold clinic. Instead, I was working in the back woods of the homestead cleaning up fallen timber and digging trenches to divert water away from the house.

This was an awkward time of year with old, dirty snow mixed with sand and gravel, and no sign of fresh greenness. Breakup. The good thing about breakup was it heralded the return of life. Daylight woke us by 7 a.m. and the mild day daytime temperatures 35° to 40° hinted relief from winter's grasp of cold and darkness. The combination of daytime warmth, which melted the snow into daytime slush, and overnight lows of 5° to 10°, which froze the slush back to ice, kept the ground in confusion — and the rest of us wading, sliding, or both.

The still-frozen ground beneath the surface did not allow the daytime water to drain properly and small lakes formed around the house. The road became a water canal. Homesteaders who did not have a gravel road would abandon attempts at wallowing with their vehicles to or from their houses; instead, they would park by a main road and slog in with boots. The Isaaks had such a situation. Their family would walk a mile back and forth to catch the school bus, to haul in groceries, and what not. It wasn't that bad for us since we were close to Gas Well Road.

"Elmer!"

I looked up and saw Ruby coming toward me, trying to walk around the waterways in her black knee-high rubber boots.

"One of your O.B.s is on the phone."

The woman had been fortunate to get through. We were on a party telephone line, shared with a number of other people. Sometimes I'd pick up the phone to make a call and hear one of the teenage girls just listening to her boyfriend breathe, or so it seemed. There had been times when I needed to make a critical call and had to break in to a conversation, introducing myself as Dr. Gaede. Obviously, there wasn't much privacy or confidentiality in my over-the-phone question-answer discernment of medical need.

"I'll be right in."

I chopped away at one more chunk of ice before turning toward the house.

Within a few minutes of telephone conversation, Mrs. Smith gave me — and anyone else who might have been listening — an experienced progress report on her condition. This was not her first baby, so without hesitation I told her I'd meet her at the clinic around 5:15 p.m.

I changed my work clothes, which were singed from brush fire burning, and headed out the door to the Volkswagen bus.

"If this is the real thing, I won't be back for supper," I called to Ruby.

The VW skated on the water-on-ice Gas Well Road to Kalifonsky Beach Road that met the Sterling Highway, and across the bridge that spanned the Kenai River. The bridge was the only one crossing the Kenai River and connected the lower Kenai Peninsula towns with the main part of the Peninsula. A state trooper recognized my vehicle and waved. The troopers expected I was on my way to a medical emergency — no matter the time of day — and I was waived from speed limits. Of course, I had to give notice of any new vehicle purchase, so I could maintain my special privileges. Ruby thought the exemption status applied to her. She found out differently.

Mrs. Smith met me in the clinic parking lot and took her muddy boots off at the door. She'd driven herself to town and was without anyone to lend support. Chances were her husband was in the oilfield and a friend was watching her children.

"It's a mess out there, isn't it?" She shook her head and held her stomach. "I didn't plan on having a baby at breakup when the roads are so bad."

She lay down on the examining table, which would most likely turn into a delivery table. I began my evaluation. Blood pressure normal. Fetal heart rate normal. The baby's head was low. I needed to call a nurse right away.

Abruptly the room swayed. I grasped the examining table to steady myself. Was I dizzy? I sat down on the nearby tall stool. The movement continued, now with a distant rumble and a stronger force. I looked at Mrs. Smith. Our puzzled eyes met. "Earthquake!"

"Let's get out of here!" I shouted above the din and helped her off the table. I held on to her arm and we careened down the hallway to the emergency ramp door, which I deemed most solid.

The shaking intensified. As we stood looking out the open door, I saw tall spruce and aspen trees whip violently back and forth until their tops nearly touched the ground. Like the sound of surf, the roar became deafening. The barn across the street jumped alive and gyrated on the convulsing ground. The ground heaved up and down like ocean waves and cars lurched crazily on the road. I'd been in earthquakes at Tanana, but never like this.

I stood horrified as a jagged crack appeared in front of a car. It opened about a foot wide and then suddenly clapped shut. The earth stretched apart and other fractures appeared. The smell of sulfur filled the air. I was staggered by the force of nature. The thunderous rolling continued and the ground groaned in agony. *Will it never end?* I wondered. *How long can this last before everything is broken apart or sucked into the earth?*

After four never-ending minutes, the nightmare stopped — or so I thought. Silence.

"I'm going home," said Mrs. Smith in a trembling voice. "I don't want to have my baby right now."

She walked into the empty waiting room, stepped into her boots, carefully made her way down the front steps and out the front door to her car.

Back in my office, the large clock on the wall, hung crookedly and had stopped at 5:36 p.m. I pushed back the furniture in the waiting room, which had danced out of place, and then tried calling Ruby. The phone was dead. I had to get home.

Just as I opened the front door a state trooper pushed in. The usually self-assured man, who dealt with terrible accidents and Alaska catastrophes, was wild-eyed and uncertain.

"Doc, you've got to stay!" His command sounded more like a plea. "Emergencies will be coming in!"

I'd never seen him so frantic and wondered what he knew that I didn't. I paused. I was in a bind between medical obligations and my concern as a father and husband, yet he'd given me no choice. I'd been ordered to stay at my post as a physician.

Later Ruby told me of her experience. She and the children were sitting at the supper table when they heard a loud thud and then felt a jolt, as though something large had run into the house. Even when they figured out it was an earthquake they expected it would subside — as earthquakes before had done. When the shaking and noise increased, she'd feared the house would crumble.

"Let's get out of here!" she'd screamed.

She and the children had made their way drunkenly toward the front door. Mishal had fallen down the steps. Ruby pulled her up. The driveway was covered with snow. Unable to maintain their balance, they'd collapsed onto the cold ground, without shoes or coats. Trees had swayed as if they were feathers. The ground had rumbled and split open, emitting swamp gas from the shallow fields beneath our homestead. After hour-long minutes, they'd returned to the house, Ruby felt nauseated and as if she'd been on a boat churning in rough seas.

The only damage she found was water sloshing out of her suds-saver tub in the laundry room and a fallen flower pot. None of her china or fragile keepsakes had tumbled out of the display case, and sugar bowls and syrup bottles in the cupboards were spared.

She'd tried to call me, but when she picked up the phone all she heard was a woman screaming hysterically. Ruby had told her that she was scared, too, but the phone lines were dead and no one could be reached. The woman remained out-of-control. Ruby hung up to deal with her own situation.

When the evening shadows crept in, she'd found candles. Remarkably, after several hours, electricity was restored and she turned on the radio — to the shocking news from a Seattle station that no one knew what had happened to Anchorage, Alaska.

In the clinic laboratory, I located a battery radio to learn about possible damage in other areas. I was surprised with the difficulty in finding stations. In the usual settings was just a lot of static. I began to wonder if anyone was "out there" or if we were completely alone. Were some areas so damaged that nothing remained? This was a big state. Certainly not all stations were down.

Finally, I tuned into a Seattle station. Gradually, and with jaw-dropping disbelief, I learned what had happened in Anchorage. The announcer's reports were so graphic and grim I couldn't comprehend the horror until I'd heard the same message over and over. Houses and people swallowed up, bridges destroyed, entire streets dropped below the surface, and fires started. The broadcasts were without music and commercials. There was no lightheartedness to break the tension. The extent of the damage in Alaska had only begun to be assessed.

At that moment, no one knew the earthquake had registered 9.2 on the Richter Scale, making it the strongest earthquake recorded in North America. Nor did they know the epicenter was 100 miles east of Anchorage — near Prince William Sound.

The Good Friday sun slipped away, edging pink wisps of clouds with gold against the darkening sky. Darkness closed around us. Hour by hour, the night grew blacker and the reports became worse. Aftershocks added to everyone's trepidation. The nightmare was not over.

A new report informed us that the earthquake had churned up a tidal wave. Our homestead was three miles from the beach and even at that distance, we were close to sea level and a quake as gigantic as we'd experienced was powerful enough to propel itself inland. In the utter blackness, no one would be able to see if it came, or have any chance of getting ahead of it.

Patients came and went during the night. I left the radio on and we listened together. I prayed and waited for the obscurity to break into daylight.

The next day I was released to go home. This was not the same town I'd driven through the day before. Signs lay crumpled on the ground, buildings had slits down their sides, and streets were cracked. I was thankful to see the bridge across the Kenai River was still intact, unlike 141 of 204 in south-central Alaska, which left many of the small towns on the Kenai Peninsula isolated.

Two days later, on Easter, the *Anchorage Daily Times* rolled out papers with preliminary lists of casualties in Anchorage and pictures of buckled downtown buildings, cars fallen into yawning pits, burst water mains, snapped power poles, and houses sloughed off the bluff down to the Cook Inlet. Governor Egan estimated damages in Anchorage alone to be at $250 million, which he said was conservative. By the time all would be assessed, calculated damage to Alaska would be over $500 million of which around 60 percent would be sustained by Anchorage. President Lyndon B. Johnson, on vacation in Texas, greeted a news conference with weary eyes after a night of receiving reports of the disaster.

Unlike Anchorage and the coastal towns, Soldotna was in pretty good shape. There was no major structural damage, and because there was no city water or sewer, no main lines were broken. Within the week we would hear cargo planes overhead bringing food supplies to Kenai.

The following day, the *Times* provided instructions for Anchorage residents regarding gasoline, food supplies, fuel oil, water and field toilets, mail delivery, typhoid shots, and schools. Casualty figures increased, although actual bodies could not be found of those swallowed up into the ground. More than 2,000 people were homeless.

Anchorage was not alone in this apocalypse. In the Portage area — a flat area along Turnagain Arm, around which the road from the Peninsula to Anchorage winds, an old log cabin was the only building in the area not condemned by Civil Defense. About 40 residents had spent the night on high ground above Portage and were evacuated by helicopters. Army engineers cut through snow and rock slides and needed to either rebuild or lay temporary bridges in 13 areas just to get to Portage, much less the entire distance to Soldotna.

This minimal damage was not the case in towns such as Homer, Kodiak, Seward, and Valdez. At Homer, only 80 miles away, the dock was ripped loose at the Spit, and boats littered the remaining waterway. The land table had dropped nearly six feet, so with high tides coming in only a few weeks, all the buildings near the dock would be flooded. The fragments of dangling dock were no longer useful at the lower elevation.

At Kodiak Island, the tidal waves heaped more damage upon earthquake destruction. Most of the boat harbor was gone and boats were strewn on the beaches. Between 650 and 700 people who had been evacuated from other parts of the island were being fed by the Civil Defense agency at the Kodiak Naval Station. Another 20 to 30 people were unaccounted for.

Reports of devastation continued. In Valdez, most of the residents were evacuated.

Governor Egan said of his hometown, "There is no sign that there ever was a dock or boat area. This area has totally disappeared."

Fires added to the chaos and 34 people were known to be dead.

The Easter church service took on a new meaning as I thought of the 104 or more people killed in the quake and the grieving of those who had lost these loved ones. I hoped they would find spiritual comfort on this day. I thought of the traditional Easter story, where an earthquake shook the enormous rock from the entrance of Jesus' tomb. The guards attending this tomb were frightened and confused — and I could certainly understand why.

I had to see for myself the bizarre turmoil resulting from the Good Friday Earthquake. Paul Isaak and I flew to Seward to view the staggering confusion there. Although Seward was closed to outsiders, we were both members of the Civil Air Patrol; furthermore, we were on the hospital staff and granted special permission to enter the area.

In reality, it didn't take much to keep people out of Seward. The road was badly broken apart, and the main portion of the runway was unusable. There was no trace of the hangar we used, and the cross runway was in shambles with heaps of gravel, trees and debris. It was a spooky feeling to realize what had been was now no more.

As if the earthquake hadn't rendered enough damage, a tidal wave had rolled in and crushed everything for about three-quarters of a mile from the bay. The mile-long waterfront had collapsed into the bay, and docks, warehouses, offices, and storage tanks had vanished. Rails, train cars, and engines were melted together or tossed about as if an angry child had tired of play. In a lagoon a half mile from Seward, two rails dipped up and down with the tide. Wrecked cars, twisted rails, and crumbled houses made what had been crowned an All American City look like a garbage dump. The smoke had so obliterated the town that originally it was reported that the entire city had been wiped out by the quake and ensuing tidal wave. It was just as the *Anchorage Daily Times* had reported,

> *The supply lifeline for the interior of Alaska — the Alaska Railroad — will have to be brought back from the near-dead. Its facilities and equipment at Seward are a mass of molten steel and burning railroad cars. Unofficial sources say it will be weeks before the 470-mile line is again in full operation.*

The eerie feeling intensified as we flew south of Seward.

"Didn't there used to be a mountain peak over there?" asked Paul.

"I thought we knew this area like the backs of our hands, but something seems different." I responded.

"Do you think an entire mountain could be swallowed up?"

I didn't answer. That concept was too overwhelming. For some time, we flew in silence.

After a while, Paul pointed out the window, "Look! That lake is empty!"

I pushed the stick forward and we flew down for a closer look.

"The bottom must have cracked open and swallowed up the water!" I couldn't believe what all we were seeing.

I felt compelled to witness Anchorage and confirm for myself the incredible radio and newspaper reports of residents living in a state of emergency. Many were without the basic necessities of sanitation, heat, and shelter. They were urged not to hoard groceries, a difficult restraint given their fear. One school had split open and consequently all schools were closed until further notice. Driving to Anchorage was impossible, and now air travel was more important than ever in this Last Frontier.

My first glimpse of Anchorage was the exclusive Turnagain Heights residential area — a previously desirable living subdivision that overlooked the Knik Arm. The ground had sloughed away and homes and cars had fallen down to the water and into gaping pits. The frozen earth was a mishmash of raised and lowered ground and stretched earth. Between 100 and 125 families had called this area home. When they'd felt their houses giving way, many had run outside. Some homes had been swallowed into the earth, which had split open and then slammed shut. Incomprehensible. At this altitude, I thought how a large hand needed to pick up the houses and push everything back together. I flew on.

Streets and lawns had been wrenched apart. A multistory apartment building lay crumpled. The walls of the new, five-story J.C. Penney building had sheared away and crushed a half dozen cars parked at the curb. Fifth Avenue was now above the previous floor level of many buildings that had sunk. The grotesque destruction went on and on.

I returned to Soldotna and tried to describe to my family what I'd witnessed. I found myself at a loss to fully communicate the terrible devastation. Buildings could be restored, and the land would eventually heal, but there would be permanent scars on children, parents, young people, men, and women as they remembered the Good Friday Earthquake.

In my line of work, death and birth were a part of the circle of life. A week after the history-making phenomena, Mrs. Smith returned and the "Earthquake Baby" *did* arrive. The child had truly arrived at "breakup" when the Alaskan world broke apart.

CHAPTER 24

BABY WON'T WAIT

September 1970

"DOC, I DON'T KNOW if it's serious, but we wonder if Barbara is starting labor. It seems her water may have broken and she's having some twinges of pain every five to ten minutes."

It was my day off. Mark and I had been working in the autumn woods, cutting and hauling in firewood, when Ruby sent Mishal out to find me.

"FAA is on the phone," my youngest announced, swatting at mosquitoes and then squatting to pick a handful of cranberries.

Chuck Crapuchettes (CRAP-a-shets) had contacted me through an FAA phone patch. He continued, "We don't know if this is just false labor or the real thing. What do you want us to do?"

Chuck taught grades four through eight at Newhalen (NOO-hale-en) School, seven miles from Iliamna (il-lee-YAM-nuh.) His wife, Barbara, managed the office as secretary and substituted when necessary. Irene Carlson, who we'd lived with in Anchorage, taught grades one through three. Our families saw each other several times a year, so when Chuck and Barbara announced the pregnancy of their fourth child, they asked me to oversee prenatal care.

I could imagine the two of them now, Chuck a congenial man with short crew-cut hair and dark-rimmed glasses, and Barbara, petite — albeit fully pregnant — with her brunette hair piled on her head, or, on a chilly day, coiled around her neck for extra warmth.

Chuck Crapuchettes

Barbara Crapuchettes

Barbara was a plucky gal and used to Bush living. Between a trip or two into my clinic in Soldotna, and a house call from me across the Cook

Inlet, the September delivery was expected to be routine. Barbara would be flown out two to three weeks before her due date and stay with Soldotna friends and then deliver her baby at the clinic.

I discussed this possible turn of events with Chuck.

"Chuck, even though Barbara isn't due for three weeks, there is the possibility this could be real labor. Is there a plane available to fly her in today?"

"No Doc. Nothing for another day or two."

"She may not make it that long," I said quickly. "I'll check with Missionary Aviation Repair Center (MARC) at the Soldotna Airport to see if they can come over right away to fly her in for possible delivery."

She had been late for the previous three pregnancies, but I wanted to be on the safe side.

"I'd better come along," I told him.

I terminated the phone call and dialed MARC. Dave Cochran, a veteran bush pilot answered. I explained the situation.

"This baby might not wait. We may have an in-air delivery."

"We have a Skymaster available — and the weather looks good." Dave responded in his usual no-need-to-worry manner. The Skymaster was a twin-engine "push-pull" aircraft with one engine on the nose and one on the back of the fuselage. The split-tailed plane was flown by MARC because of the reliability of the two engines, the good performance on short airstrips, and the relative economy.

Barbara's speedy deliveries were the only commonality among her one-of-a-kind birthing experiences. I had a hunch this one would add to the mini-dramas of the others.

When she was pregnant with her first child, Barbara had been told that a first labor would be ten to twelve hours. She and Chuck had been living in Naknek where Chuck was doing commercial fishing in Bristol Bay. The nearest hospital was in Dillingham, 60 miles away and across Kvichak (KWEE-jak) Bay. The hospital was for Natives only, but a private doctor, Dr. John Libby, had agreed to deliver her. Even though that agreement was in place, when Barbara had gone into labor early, he was out of the area. At near midnight, Chuck ran around the village trying to find one of three midwives in Naknek. Even though it was summertime in Alaska, the August sun had retreated for the night and people were in bed. He knocked on doors of darkened cabins. No luck. Two midwives were at fish camp, and the other was stone drunk.

Chuck and Barbara were staying in the little Arctic Missions house, and in Chuck's franticness, he'd accidentally uncovered two small birthing manuals put out by the Alaska Department of Health: *When Baby is Born at Home,* and *Manual for Alaska's Midwives.* The booklets contained more hand-sketched pictures than written instructions.

Although these parents-to-be were stalwart Alaskans, these crudely drawn manuals provided more information than either had on their own. Doing some speed-reading and intuitive deciphering of the diagrams, the couple teamed together for the delivery. Shortly thereafter, Amiel entered the world in a manner fitting for that of a Last Frontier child — alone with her parents, in a simple and dimly-lit bedroom, by the hand of her father.

A second child, Mark, was born in a drastically different, and in contrast, mundane environment — a hospital in Anchorage.

The third child, Joel, had a story worth telling. By this time, the Crapuchettes had moved to Iliamna — kind of. Most of the summer Chuck was fishing in Bristol Bay while Barbara and the two children were staying in a boarded up, dilapidated building with a dirt basement. In the 1930s, it had been a boarding house with a completely open upstairs for bunks.

Barbara had her hands full, but took everything in stride. First, she started a rustic fast-food business and sold hamburgers to guests dropping into the area for fishing and hunting. Then, for a half-dozen guys in a surveying crew, she flipped pancakes and sent them on their way with a sack lunch each day for three weeks. She named the base of her operation "Iliaska Lodge," blending together "Iliamna" and "Alaska." (In the future, this lodge would house Tanalian Bible Camp.)

Cooking and tending children weren't her only duties. There was the light plant to manage, which supplied electricity and operated the pump that pulled water from Iliamna Lake to barrels on top of the lodge roof.

As is typical of many Bush and homestead dirt airstrips, the airfield in Iliamna was semi-jelled during spring break-up. No planes landed for nearly six weeks, the amount of time it takes for winter ground frost to finally melt and allow the water on top to drain. Such had been the case in April 1967.

Barbara had caught a plane ride to Soldotna prior her due date and stayed at our house. The magic date came and went. After ten days, I decided to induce her. After three hours and twenty minutes of labor, Baby Joel had come squalling into the world.

Barbara had been away from her family for over a month. Even though on-shore landing was not an option, she wanted to get back as soon as possible. In 1964, shortly after we had moved to the homestead, I'd sold the PA-14 Family Cruiser and purchased a Maule Rocket. On this occasion, I still had skis on my new airplane and I figured there would still be some good ice on Lake Iliamna. Although there were no phones in Iliamna, and Chuck had no idea we were flying in, I was expecting that when I buzzed the village, Chuck could figure out that I wanted to land, with his wife and new son, on the semi-frozen lake ice and would run out to indicate where it was safe to touch down.

I roared over the lodge, but no one emerged. I knew much of the ice was rotten, especially beyond pressure ridges, and I could easily crack through.

"Barbara, get ready to jump out!" I urged.

She clutched her small bundle and pulled on the door latch.

"I'll throw your suitcase out behind you."

No one came running to point me to solid landing. I was on my own. As I touched down, I nervously watched for signs of the ice giving way. I neared the shore apprehensively and slowed the plane to a near stop.

"Jump!" I bellowed.

The bush mama burst out the door. I pushed the suitcase after her, then advanced the throttle to keep the plane moving over the weak ice. Carefully I accelerated to take-off speed.

Before leaving the area, I circled back to where mother and child were picking their way carefully over the rough, yet slippery ice, toward the shoreline. Chuck walked rapidly toward them. The proud parents each waved. The suitcase lay on its side, no contents spilled.

Now, I was preparing for baby number four — and not sure how things would turn out. Perhaps it *was* false labor. Perhaps Dave and I *would* get her back to Soldotna in time for a clinic delivery. As far as I knew, there were no midwives in the village. All the Natives went to the Anchorage Native Service Hospital to have their babies, and, not just the pregnant mother, but the entire family would leave six weeks prior to the due date, and remain in Anchorage for six weeks following.

The weather was partly cloudy and calm. I'd flown with Dave many times before and our similar-aged children spent time together. I enjoyed our conversation as we winged our way through Lake Clark Pass, noting the changing colors on the hillsides. The Alaska autumn was all too short. About 20 minutes out, Dave called the Iliamna FAA to confirm our estimated time of arrival. He instructed them to contact our passengers for immediate departure.

Sure enough, Chuck and Barbara stood waiting when we landed. Dave stayed in the plane in readiness to make a swift departure, but I climbed out and greeted the parents-to-be. Seeing Barbara clutch her stomach, I felt uneasy.

"I think I'd better examine you before we load up. It's a bit cramped inside there to deliver a baby."

I looked around and saw a Jeep near the FAA station.

"Let's see if we can borrow the back of that Jeep. Is there a blanket anywhere?"

Within minutes I exclaimed, "Barbara, you're not going anywhere! You're six centimeters dilated and will probably have that baby within the hour."

The Crapuchettes family no longer lived in Iliaska Lodge, but had moved into the Newhalen School teacherage. Chuck, who was unperturbed by the change of plans, said, "We may as well go back home to our bedroom."

Dave tied down the Skymaster and stayed with the plane. I reached for my emergency delivery pack in the backseat of the plane and the three of us caught a ride to the teacherage. Word spread quickly, and on our heels, an elated string of Natives followed. It seemed the entire village was coming for the event. This would be the very first baby to be born in the village and everyone wanted to witness the drama.

Barbara and I went into a bedroom. People edged through the front door in a steady stream. Chuck put on the teakettle. The front door squeaked with new arrivals. Tea was served. Excited chatter filled the background. Occasionally I'd glance around to see a stack of eyes peering curiously through the gap of the narrowly opened bedroom door.

Chuck pushed through to check on the progress. "There are about 35 to 40 people wanting to join in."

Before I could comment, the baby's head popped out and the newborn gave a lusty cry. Seven-year-old sister, Amiel, heard the sound of a new sibling and ran in just as the rest of the baby emerged. She stared wide-eyed and speechless.

"I guess you won't have to tell her where baby's come from," I remarked. Then, turning to the blonde-headed youngster, I continued, "You have a little sister now."

She stood for a moment, and then burst out the door to make the announcement to the celebrative crowd. The well-wishers cheered and clapped. Amiel returned beaming, with her brother, Mark, in tow to show him tiny Sara.

All seemed to be well and within short order I threaded my way through the happy congestion and out the door. One of the men gave me a ride to the airstrip.

Halfway back over the inlet, I turned to Dave, "Well, this is about where Barb's baby would have been born."

"Better there than here," he chuckled.

I agreed. It would have been tight.

Bush men and women need bush doctors. Bush people need other bush people. We were a pioneering bunch, trail-breakers in most everything, including bringing new life into the Alaska wilderness.

CHAPTER 25

RETURN TO POINT HOPE

Fourth of July Weekend, 1982

THE WIND BLEW STEADILY in our faces and our feet clung to the sand as if they were in wet cement. White-capped waves pounded against the beach, sending frothy fingers of water our way. We veered up toward the short grass-tufted bluff. Fog ebbed and flowed, teasing us with mirages of people and houses just beyond the next tundra knoll. Chilled to the bone, it took too much effort to talk, and we trudged wordlessly against the resistance of the elements.

This arctic coastline beach was familiar. I'd been forced down here twice before: once after the polar bear hunt at Point Hope, and then on my way back from Barrow. I'd flown over one other time — and had managed to stay off the beach and in the air. Over the years, some things hadn't changed, such as the wind, which meant business, and cut through my insulated jacket, blue plaid flannel shirt, and T-shirt. I pressed through the thick, blowing mist.

I wasn't alone. Mishal, now a young woman of 24, walked beside me. Her long, curly black hair tried to free itself from her lavender knit hat. Ruth's husband, Roger, bent his head into the wind and pulled his tan, hooded jacket around his lean body. He could have passed for my own son with his same five-foot ten-inch height and dark hair. My sister, Lillian Pauls, added a fourth to our small expedition. She was from California and had anticipated an interesting Alaska sight-seeing trip. "Rugged" or "daring" would not be words to describe her.

This was not the first time Mishal would be flying to Pt. Hope with me.

My family had grown up. Naomi, a writer and speaker with a degree in English Education, had married Bryan Penner, a construction engineer who had started his own general contracting company, Maranatha Construction, as well as acquired a private pilot's license — and flew a Cessna 210. They lived in Denver, Colorado, with their two children.

Ruth had obtained her Licensed Practical Nurse degree, had horses, loved cats, and spent hours in her garden. She had married Roger Rupp. He was an Alaskan kid, too. His parents had been schoolteachers along the Yukon River at Beaver and on the Tanana River at Tanacross. Later they were houseparents at the Lazy Mountain Children's Home near Palmer, Alaska. Roger had graduated from Moody Bible Institute and was a pilot, flight instructor, an A&P (Airframe and Power Plant) mechanic, and authorized inspector. He and Ruth lived across the homestead airstrip, where they'd built a house for their family of three children.

Roger conveniently conducted Roger Rupp Aircraft Service in his nearby hangar. On the airstrip in front of their house, a multicolored lineup of his customer's planes added a vibrant hue, especially in the winter. Having Roger nearby was like having my own personal mechanic, and given my flying history, everyone knew what a benefit *that* was.

A long stone's throw from Ruth and Roger's place, smoke floated out the chimney of the cedar house Mark had built. He'd married Patti Kvalvik, a local girl who'd lived just up Kalifornsky Beach Road. The same year we'd moved to Soldotna, her father had been offered a job with Homer Electric Association (HEA.) He had driven the family Rambler up the Alcan from Sand Point, Idaho. Her mother had flown up several weeks later, and landed in Kenai on Wien Airlines with three children.

Although Mark had spent plenty of time drawing blood and looking through a microscope in the clinic laboratory, he'd become a computer technician for Dresser Atlas, an oil industry service company. Patti was a bookkeeper for Soldotna Drugstore, a pharmacy and variety store. They had one daughter.

Mark used the airstrip, too. Ever since he'd been born, he'd been my flying buddy, sitting on a sleeping bag, peering out the window, and ready to grab the stick. As he got older, I let him have the stick — and gradually more and more of the controls. He spent time in my fleet, from the first J-3 to the Family Cruiser; next, the Maule, followed by another J-3, a Piper PA-12 Super Cruiser, a Piper PA-18 Super Cub, and finally the Cessna 180. Flying these airplanes came as easily to him as riding a bicycle. He'd never gone from "Today I am going to start learning to fly," to "Now I am a pilot." Early on, he could land and take off. It was more like, "Today I am officially going to start on my pilot's license" — which meant learning the regulations, acquiring the technical skills to keep from inadvertently doing something stupid, and accumulating hours in a log book to qualify for a check ride. At age 16, he soloed with a laugh and a grin; he'd been flying unaccompanied before this, but now he was legal. Already, he'd started his own flying adventures.

For a while, Mishal migrated to a Denver Art School, off and on living with Naomi, and then returning to the homestead each summer. One day when she was reading the *Anchorage Daily Times,* a photo of an Inupiat Eskimo girl from Point Hope caught her eye. The girl was visiting Sea

World in San Diego. Beneath the picture was the name, "Darlene Tooyak (TOY-yuck)." We'd told Mishal that her birth mother was Dora Tooyak. She knew this girl must be her relative.

When Mishal shared this news, I wasn't disturbed, but Ruby had been. She wondered if Mishal would regret being adopted into our family and emotionally detach from us. Mishal had her own mix of emotions.

With trepidation and hope she contacted the Tooyaks in Point Hope. Of all things, she learned her birth mother lived in Denver! Now she had the opportunity to meet the woman she had wondered about all these years — yet, did she really want to? Apprehensively and thoughtfully, Mishal arranged a meeting. The reunion, a mix of surprise and bittersweetness, went well, and was followed by an invitation from the Tooyak relatives in Point Hope, the following spring of 1981. It wasn't long before I made plans to fly her up to see her place and people of origin.

Mishal had driven up the Alcan with a friend, Sue Erickson, who she'd met through Naomi. In contrast to Mishal, who was short, strongly built, and brown-eyed, Sue was of Scandinavian ancestry and tall, blond, and blue-eyed. This new friend fit right into our family. She was a veterinary assistant and horsewoman, so she connected immediately with Ruth.

I hadn't preannounced the trip to Mishal, so it wasn't until the two friends arrived that either knew they'd be flying out for a memorable Memorial Day weekend. When I'd first met Sue, I could tell immediately she had a sense of adventure, and now she didn't hesitate a minute when I invited her for a cross-country trip to Point Hope in the 180 Cessna.

Ruby, however, worried. She'd written to my brother Harold and his wife Marianna:

"...Friday noon Elmer, Mishal, and Sue left for Pt. Hope. They were supposed to have good weather all the way. They gassed up once and should

have gotten there for night. I wish Elmer would call me on trips like these but he files flight plans so someone keeps track of him. I'm sure by now they have a phone there. 22 years ago when he went Polar Bear hunting they didn't. He didn't even file flight plans then."

She had no cause for concern. Our trip was routine and without incident. We circled the village and landed on the beach — and nearly got stuck in the deep, loose gravel. The local policeman met us on his red Honda 110 three-wheeler, and within short order, other villagers surrounded us with their similar vehicles.

Mishal was joyously welcomed by her blood relatives in the village of 300 people. Her uncle, Andrew Tooyak, who was mayor, invited us to stay with him and his wife, Irene, and we were treated royally. Irene's disposition was as cheerful as the never setting sun. Andrew showed us his ivory carvings of an eagle, polar bear, and a watchband with tiny Alaska animals; all these items demonstrated his delicate and careful craftsmanship.

On one occasion, we were in a house crowded with Native friends and relatives. The small room was a hubbub of conversations. While Andrew and I talked, I could see over his shoulder and observe Mishal and Sue raptly engaged with two Inupiat women, still in their parkas, but with hoods pushed back and long hair flowing around their necks. They talked animatedly in Inupiaq, then switched into English, and then back into Inupiaq; the exchange was accompanied with gestures. All at once, a toddler appeared from nowhere and sat sprawled across Sue's lap! Apparently, the child had been asleep inside his mother's parka, and when awakened, crawled out to what was nearest — Sue's lap.

At this time, the end of May, there were three-to-four-foot snow drifts still present, but instead of using snow-machines, we used three-wheelers to get around. It seemed every man, woman, and child owned at least one.

While we were there, Andrew's sons, Andrew Jr. and Enoch, shot two walruses and several seals. Sue was certainly getting a true Alaska experience and wasn't finicky when it came to tasting walrus and caribou meat.

Mishal never stopped beaming and appeared easily at home with everyone. When it came time to leave, her relatives and other villagers embraced her, and asked repeatedly, "When will we see you again?" I trusted this visit would be the start of a satisfying journey in learning about her rich heritage. The fulfillment was deepened by a stop at Tanana on our way home. The Tanana people remembered Mishal's mother, Dora, and Mishal was honored by their open-arm and open-heart reception.

The next leg of our trip was altered due to forest fires near Fairbanks. We had to fly off course to skirt the area, however, I was not disturbed by this change in flight plan. That is, I was not disturbed until the engine began running rough. It would smooth out and I'd relax, but then a rattle would shake me out of my complacency. The occasional puffs of smoke coming from it *did* cause alarm! I'd had my share of airplane problems, but never anything like this. I didn't deny the reality to Mishal and Sue, "This is bad. We're really in trouble if this plane catches fire. We won't get down alive."

The Cessna 180 shuddered. I got on the radio to Fairbank's FAA to give our location. They gave me directions to Healy, about 20 minutes away. I instructed my passengers to look for any emergency landing spot.

"Any place I can land, a sandbar, along the river, any clearing."

Nothing but thick spruce.

Mishal put her face near to my ear, "Dad, I think we'd better say 'goodbye' to each other — just in case we don't make it."

I looked at the face of my youngest child. She was a beautiful woman with so much to live for, and she was pretty grown-up to show this maturity.

"I love you Dad," she said. Her eyes teared up, but she remained steady.

"I love you, too, Mishal."

She turned to Sue in the backseat. I couldn't see her face, but I heard her say, "Sue, I'm sorry it has to end this way."

I winced as again the plane shook. Smoke wisps blew out of the engine and past my window

Fairbank's radio informed me they had passed my whereabouts on to pilots in the area. "They've called for your location and are getting into the air now."

One way or another, I'd have assistance, whether that was to spot where I'd crashed and/or to rescue us *if* we survived.

Except for my frequent calls to Fairbanks FAA, the cabin was silent, with an odd sense of peace.

We managed to stay in the air. No flames — yet. After an agonizing quarter hour, we got to Healy, on the Parks Highway between Fairbanks and Anchorage. The dirt strip was no bigger or sophisticated than that on the homestead, and there was no radio control tower. I contacted Fairbanks and told them we'd be attempting a landing. I prepared my passengers for the worst.

"I don't know if the plane's controls will respond to a landing, or if the engine will burst into flames as we descend."

Carefully, and with inch-by-inch maneuvering, I brought the plane down. The wheels touched the earth. We still weren't out of danger. We rolled to a stop and I shut down the engine.

"Get out now!" I yelled, and pushed across Mishal to open the door. We bolted and ran as hard as we could, expecting an explosion at any moment.

Finally, a good distance from the plane, we slowed to a breathless stagger and turned to see if the plane was on fire. The air was still and the sun softened by the haze of the wildfire smoke. Birds chirped. We watched and waited. Our labored panting and anxiety seemed out of place against the peaceful setting.

A bearded man in his 30s walked toward us rapidly. His heavy blue and black plaid jacket blew open as he broke into a run, and his dark hair ruffled off his forehead.

"You okay? I didn't think you'd make it down!" He puffed from exertion and concern. The man, Larry Miller, was a bush pilot himself and had ditched a plane or two. He knew a smoking airplane spelled doom.

Cautiously, we walked to the 180. The belly was coated with oil and when I lifted the cowling and checked the dipstick, not a drop of oil remained. We gaped wordlessly at one another. Either we'd had amazing luck or we'd experienced a miracle.

This luck, or miracle, didn't stop with the safe landing. Larry had *happened* to be at that nothing-of-an-airstrip on Memorial Day to fix a thing or two on his own airplane, which would only take a short time — which happened to be the same time we found our way down. And, if we'd crashed, the guy was an EMT. As are the likes of Alaska folk, Larry took us home in his super-cab pickup so I could call Roger to rescue us. Larry's wife, Debbie, made us sandwiches and tea, which we savored with a view of their lake. The tranquil setting denied the reality of the morning's near-death experience.

Late afternoon, Roger and Dick Page, a mechanic and pilot for MARC, flew in with MARC's Cessna 310.

"Are you ready to climb into another airplane?" I said to Sue with a wink.

She answered by climbing in.

The low-wing, twin-engine aircraft cruised past Mount McKinley and brought us home, concluding our Memorial Day weekend — *without* us making our own memorials.

A week later, I wrote my mom in Reedley, California:

> *"I called Roger to come out and pick us up. We were about 250 miles from home. We checked out the engine, five of the six pistons and cylinders were faulty. Rings were broken and cylinders deeply grooved. I'd just had the engine overhauled in Anchorage before this trip, so we believe the shop in Anchorage must have put in faulty parts..."*

Roger's investigation revealed that the recently overhauled engine had not been broken in correctly. The ignition timing had been set far too advanced, causing the engine to run hot and wear excessively.

Now, July 1, 1982, Mishal and I were returning to Point Hope, this time with Roger and Lillian. When we'd left, we hadn't intended to take a walking tour of the coastline, nor had I expected to set down another plane on this beach. Mishal had insisted Roger fly with us as co-pilot and on-hand mechanic.

"Dad, in case we have plane problems, she'd said firmly and unbendingly.

Her memory of my flying mishaps remained fresh in her mind.

My sister, Lillian, visiting from Reedley, California, was eager to join in. When we'd walked out the door to my hangar, I'd overheard Ruby tell Lillian, "I hope you have fun. Long trips with Elmer always turn into some kind of adventure."

Roger and I rolled the six-place 180 out of the hanger and the four of us climbed in. Behind the second set of seats was a bench, suitable for kids only. We taxied to the homestead airstrip, and after checking the vicinity for Ruth's horses, I spun the plane around in front of Roger's hangar. My able co-pilot watched and listened as I completed the flight checklist. The plane was heavily loaded with emergency gear — a bit more than my early years of a can of Spam, two chocolate bars, a book of matches, and airplane gear. Mishal had prepared a gift basket for her Aunt Irene Tooyak, and Lillian had packed home-dried fruit she'd brought from California. Roger had gathered up a large hammer, pipe wrench, crescent wrench, small anvil, and a hack saw. Notwithstanding the extra weight, by the time we passed Mark's house we were airborne.

The sky was variegated gray, but visibility was adequate. I pointed the plane's nose north and crossed the Cook Inlet. Oil platforms stood on their long legs in the deep water. We flew over the small village of Tyonek (ty-O-neck) and then skirted the Anchorage mudflats. Rainy Pass in the Alaska Range would be our first challenge.

All mountain passes stir up weather evils since fog, clouds, and snow get stuck in the higher elevations. Winds, which funnel through the passes, intensify the problem. I'd flown through Rainy Pass a number of times. It was the trickiest and most dangerous pass I'd ever been through. Today was no exception. The hostile air had been waiting for us.

"Seat belts?" I called over my shoulder.

Updrafts threatened to throw us around the cabin, as well as everything that was not securely fastened. The air currents yanked us wildly higher and higher. Then, the bottom dropped out with a downdraft. Flying along in this fashion was much like riding a bucking bronco. Near the middle of the pass, light snow blew across the windshield, but fortunately didn't pose a threat. In the distance, I could see light at the end of the tunnel. Bright sunshine beckoned us to flatter country and calmer air.

Having escaped the pass, the flight continued over winding rivers and nameless lakes. I dipped a wing and pointed out a section of the Iditarod trail. The next major landmark came into view, the wide slate-gray Yukon River, which we followed west. Bleached driftwood decorated the shoreline.

We made our first stop at Galena. The village on the north bank of the Yukon had been altered since we'd lived in Tanana. Due to a severe flood in 1971, a new community site was developed at Alexander Lake, one-and-a-half miles east of the original town site. City offices, a health clinic, schools, a laundromat, store, and more than 150 homes were constructed at "New Town." (In 2013, the entire area would flood when the Yukon River jammed during breakup.)

At this time, the Air Force Station was still in operation, although it would close in 1993.

We stretched our legs and I fueled the plane at two dollars per gallon. In contrast to the 20-some gallons fuel capacity of the J-3 I'd flown on my first trip to Point Hope, the 180 carried 88 gallons. Flying with the little 75 hp craft had taken forever; either getting batted around by weather or constantly refueling. Now with a 230 hp engine and increased airspeed of 141 knots, these cross-country trips required much less time — weather permitting.

Back in the sky, the air was smooth. Below, the treeless terrain of tan-brown tundra and dimples of small lakes might have seemed mundane to some observers, but after spending two years in Tanana with frequent flights in this area, it felt familiar and comfortable. After a while, the cheerful sunshine left us and grumpy clouds bunched together to escort us towards Kotzebue. The weather continued to worsen and drizzle steadily ran off the windshield.

"I don't mind flying in rain," I commented to Roger. "But I don't care to encounter the nasty fog along the coastline. I'll see what FAA has to report."

I reached for the radio controls.

After a few moments, Roger looked at me expectantly.

"Kivalina is socked in. We'll have to wait it out in Kotzebue."

Over my shoulder I informed Mishal and Lillian, "We're going to do some sightseeing."

I put the plane's nose down and brought it in for a landing.

Old and new culture overlapped throughout Kotzebue, the second largest Inupiat Eskimo village in Alaska. Now in mid-summer it was a busy place of sights and sounds. Caribou racks and skins hung on a house next to a Dairy Queen. A dogsled leaned against a snow-machine. In front of us, a tour bus lurched to a stop and exuberant tourists gingerly tumbled out into the wind and rain, then hopped about trying to avoid the end-to-end puddles. Seagulls argued and fought. A bulky Air Force cargo plane lumbered through the skies and other airplanes landed and took off. The unmistakable odor of fish brought our attention to drying racks. Three-wheelers with drivers of all ages careened down the street, splashing water everywhere. Some carried entire families, with children hanging on wherever possible, limbs waving here and there. Non-working ATVs and broken-down snow-machines were left half-buried on the beach.

We walked along the sloppy streets to the radio station, where we could send a message by *Tundra Tom Toms* radio, to Mishal's uncle Andrew Tooyak, explaining our delay. It wouldn't matter if he heard it himself; villagers always passed on to one another any information pertinent to a specific individual.

After communicating our situation via radio station, we squeezed in among the tourists at the jade factory. The jade was cut from the jade hills to the east along the river. After we observed jade jewelry before and after polishing, we realized we'd walked past enormous boulders of raw jade on our way from the airport; it wasn't as if someone could just pick up the treasures and sneak them into his or her pocket.

The dreary skies lingered, and there was no urgency to take off. Mishal knew her biological aunt, Viola Norton, lived in Kotzebue, but had never met her. She inquired at the local stores until she was pointed to the residential area, a couple blocks from the beach. We found the house and Mishal knocked on the door. The rest of us stood back. A woman answered the door hesitantly and we could overhear Mishal trying to verify her identity, that of Dora Tooyak's daughter. Viola didn't look convinced. She glanced over Mishal's shoulder at the three of us. I smiled and raised my hand in a gesture of friendliness. After more conversation, Mishal turned and motioned for us to join her.

A man appeared at the door and Mishal introduced him as Violet's husband Cyrus. The couple invited us in and Viola put a teapot on the stove. Undisturbed by our presence, their two teenagers slept on the floor.

In the summertime, when the sun shone nearly endlessly, the Natives tended to go until they dropped. No one wanted to waste the daylight.

Between the discovery that Mishal really was their relative and Lillian's questions, the conversation flowed and we appreciated the warm beverages and cookies. An hour later, Cyrus took us to the garage.

"Viola and I speared these a few days ago," he said pointing to the heads and tusks of three walruses. "We were up by Kivalina in our fishing boat."

This was fascinating information. All the same, I glanced out the window at the skies. I hoped by now the fog had cleared out of Kivalina. If we could make it that far, we'd have a good foothold in our trek up to Point Hope.

Elmer resting on the Kotzebue beach, 1982

At the Kotzebue airport, I ran into a local pilot fueling his plane. Trying to sound nonchalant I asked, "How are the beaches for landing?"

"If they're wet, they're okay, but if they're dry watch out," he advised.

I was ready to try it. After three hours of touring Kotzebue and waiting out the weather, we climbed back into the plane. The coastal Eskimo village of Kivalina was clear. We weren't that far from our destination now.

I'd thought we were home-free, but a short distance past this landmark, thick clouds concealed the ground, affording us an infrequent peak at the tundra and lakes below and squelching our hopes.

"Check to see if there are any other pilots around," suggested Roger.

I got on the radio and requested a weather report from any pilot who might be flying in the area.

"Point Hope is socked in with fog. Can't find a hole to get down. I'm turning back," a pilot called back.

I didn't want to turn back, so without divulging the report to Mishal or Lillian, I leaned toward Roger.

"See if you can find a hole."

Toward the north, Roger spotted a break in the clouds. I flew to it, and spiraled down until I was beneath the white blanket, then scooted over the shoreline at 200 to 300 feet. I wasn't worried now. I recognized the landmarks and estimated we were only 15 miles from Point Hope.

As we approached Point Hope's airfield, the fog rapidly crowded us closer to the sand. Within seconds our forward visibility was totally obscured. I made a rapid 180-degree turn, tried to maintain some altitude, and backtracked, hoping to find the hole we'd come down through. No luck. Mist from the open water joined the fog and thwarted my efforts. There was no place to escape.

"I'm putting it down," I said tersely to Roger. "The beach is wide and smooth, and goes for miles. It's good enough for a DC-3 to land."

Without taking my eyes off the beach, I shouted to Lillian and Mishal, "Hang on!"

With first-notch flaps, we slowly settled onto the inviting sand. To my amazement, the landing was perfectly smooth. My relief, however, was premature. The plane decelerated abruptly. At first, I thought my brakes had locked. Then I realized the sand was too dry and soft. Looking out my window, I saw the tires were plowing six-inch furrows. Using full power, I tried to force the bogging wheels up the bank to higher beach. No use. The plane relinquished its struggle with the warned-about dry sand.

"This is the end of the line for now," I said, shutting down the engine.

Roger and I scouted the area for potential firm ground and a suitable takeoff strip. Several 100 feet away, and at the very top of the beach, the ground no longer sunk beneath our feet. A section of 250 feet was suitable for takeoff. We just needed to get the plane to higher ground. This was a problem. The plane's engine could not propel the plane from where it was stuck and there was no way we four could push or pull it ourselves.

"We need a solid tracking surface," I concluded.

Like beach scavengers, we searched for pieces of flat driftwood and scrapped water-stained plywood to place under the wheels. The activity at least distracted us from the cold. Following much huffing and puffing, we pushed the plane along the makeshift track to higher ground. With a tight smile I affirmed our efforts.

This was all a step in the right direction, but I needed to reposition the plane a bit farther along the beach for our takeoff. In consideration of the added weight of passengers, I decided to do this myself. Roger, Mishal, and Lillian stood beside the pile of emergency gear we'd unloaded to further reduce the weight.

The engine started easily and I felt the tires slowly moving along the sand. My confidence was short lived. Unfortunately, I had a 15 to 20 mph tailwind and before I knew what was happening, the front wheels dropped into a soft spot, and a gust caught the wings. The plane tail shot up into the air, forcing the rotating prop into the sand. Immediately, I turned off the ignition. After a stunned moment, I had the where-with-all to find my tie-down ropes in the awkwardly angled plane. Fuel poured out of the wing tanks. I pushed open the door and crawled out, managing to dodge the stream of gasoline.

Cessna 180 on its nose, near Pt. Hope, 1982

Roger ran toward the plane as fast as he could in the ocean gale and tripping sand. I lassoed the tail wheel and we pulled the tail back down together.

Stating the obvious, I said, "I guess we'd better check out damage to the prop." I really did not want to see the bad news.

Roger examined the prop, running his fingers along the small two to three-inch curls at the prop ends.

"This plane isn't going anywhere until we can fix this prop."

I knew it, but didn't want to hear or believe it. We were really stuck now.

Lillian and Mishal stood hugging themselves, shivering and mute.

"Well gang, it looks as if the plane will stay here," I said grimly. "We can either wait until someone finds us, or we can walk to the village."

The first alternative was not really an option. No one even knew we were out here. I looked at my watch. 8:30 p.m.

"How far away do you think we are?" asked Mishal.

"I'd guess about five or six miles," I replied.

A strong wind cut through the damp fog, and I estimated the temperature to be around 42°. Before leaving the plane, we pulled on any extra clothes we'd brought along. Not much. I'd told everyone to bring only one change of dress. This would not be a leisurely stroll on a sun-kissed California beach, especially in the early evening which was already darkened by the moisture-thick storm. Rather, it would be a real nature hike. Instead of a bird's-eye view of this northland, we would get an up-close wildlife view. There *was* sightseeing. Skeletal parts and skin of whales and caribou littered the beach, along with driftwood, and we almost stumbled over a dead walrus.

We could all have used heavier coats and winter gear. At least Roger, Mishal, and I had on leather hiking boots. Lillian was the least prepared for these conditions and not as used to the uneven walking terrain as were the three of us. The minutes turned into hours, and I found myself shaking from the cold and the wetness that had seeped down to my marrow. Lillian wanted to slow down or at least stumble at her own pace, but I didn't dare let our small group lose sight of one another. Disorientation, hypothermia, and exhaustion were our enemies, more so than wild animals. The wind cut to my bones. A scrap of plastic flapped under a piece of driftwood and I wrapped it around me for protection against the wind. I needed to walk harder to stay warm, but I didn't want Lillian to straggle behind.

"Look! Three-wheeler tracks!" Roger called out above the gale.

Civilization had to be nearby. We squinted against the ceaseless murkiness. Nothing appeared. Even though the end-of-day sun only hovered at the horizon at this time of year, the fog prevented much light to filter through.

After a while, ice and snow on the beach compelled us to walk above the bank on the spongy tundra. Hardened snow and ice mixed with blowing wet grass. Our pant legs became soaked to the knees and clung to our legs. At first, we talked, but then our energy went solely into walking. Up and down the hummocks we trudged, always straining our eyes for sight of the village. Sea spray fogged up our glasses. I tried to rub them clear to see my watch. We'd walked for six-and-a-half hours. We desperately needed to get out of the damp cold in order to maintain our body heat. Surely the village would soon appear.

Mishal walked ahead of Lillian and me. All of a sudden, she shouted, "Look at the snow-machine!" She pointed at an abandoned contraption. "And here's a real path."

A few minutes later, Roger who was walking with her, stopped,"Listen. I think I hear fireworks and three-wheelers."

We quickened our pace.

"There's the orange runway marker," he called back elatedly.

"The runway is about a mile long," I said to my weary troop."And the village is a mile-and-a-half after that. We're going to make it."

I was speaking to myself as much as to anyone else. We had to rally and keep going. At different paces we faltered forward.

At approximately 3:30 a.m., we left the tundra and walked into the graveled village of Point Hope. The 40 or so prefabricated houses on short stilts all looked the same, except for some variety in the earth-tone colors. None of them had names or house numbers, and we weren't sure where Mishal's uncle Andrew lived. Although social calls aren't usually made at that hour of the night, we looked around for houses with inside lights on. We were like trick-or-treaters knocking on the doors in the ghostly cobwebbed fog. No one answered. Lillian and I tried public buildings, and then the empty Episcopal Church. The door creaked open and we entered the sanctuary. Not much heat, but it was dry, and out of the taunting wind. I poked around and found a goodwill box with a coat for Lillian. She was still trying to get comfortable when I fell asleep.

Meanwhile, Roger and Mishal were determined to find her uncle's house. Rounding the corner of a building, they were almost run over by a pickup truck, unexpected at that time of early morning and unforeseen in the dimness. Mishal explained the odd circumstances of their arrival, and the driver took them to her Uncle Andrew's house. The driver, a distant step-cousin, explained that Andrew and Irene were at the beach, at their fish camp; regardless, they could stay at the house. Roger and Mishal urged the driver to find Lillian and me, then tumbled into beds and immediately fell asleep.

The Good Samaritan searched until he found us, startling Lillian with his big husky who burst through the church doors and sniffed her in the face.

"Is anyone in here!" the man's voice boomed in the shadowy silence.

That got my attention and I groggily explained who we were, which matched the description Roger and Mishal had given. His loud entry had nothing to do with his hospitableness, and he insisted we stay at his house. Now I could completely give in to my fatigue. I embraced sleep gratefully.

Brilliant sunshine woke us the next morning, wiping away the fog that had enshrouded the village and seacoast the night before. It was as though we had gone to bed on one planet and awakened on another. We could see clearly for miles around.

Roger, Mishal, and I were unexpectedly rejuvenated and set out to find Uncle Andrew and Aunt Irene. Lillian stayed behind, utterly exhausted. This time, only a gentle breeze blew in our faces as we walked along the beach. Beluga whales lazily roped up and down through the water. We smiled at one another, making short comments of pride for our endurance, humility that God had saved us, and satisfaction that we'd survived our yesterday's "Man and Woman against Nature" ordeal.

Just as 24 years before, when I'd polar bear hunted here, the villagers rallied to our need. Andrew's son, Enoch, supplied his fishing boat and took Roger and me back to the airplane. Now we'd be using all the tools my son-in-law had sensibly and intuitively brought along. Thankfully, he knew exactly what needed to be done.

We reached the shore where the disabled plane waited. From there we looked across the water at the village, precariously balanced on the spit of land jutting into the ocean. We had been so close, yet so far away as we walked for nearly 13 miles following the curve of the land to the village.

After completely unloading the plane to make it as light as possible, and piling everything in Enoch's boat, Roger worked on the propeller and examined the engine for further damage. During this time, about 12 other Eskimos arrived in fishing boats. Together, we walked the plane to the takeoff area. In the process, it once again flipped forward on its nose. A second time, Roger worked on the propeller.

I'd pulled this plane out of some pretty tight spots before, but the useable takeoff area now was very, very short, followed by the capricious sand which was waiting to trip me up. Even if I conquered both of these, the open ocean came next. There was no margin of error. Not only was the landing surface in question, but the pilot needed every ounce of savvy to deal with the unexpected.

"Roger, how would you like the honor of flying the plane to the village?" I asked, thinking this would be logical. "You are about 15 pounds lighter than I am and besides, you've had more flying experience than me."

Roger didn't bat an eye, "It's your plane, and I'll let you take the big chance." He instructed me deliberately, "Use first flaps and get past the first 200 feet. You'll be light enough to ride the top of the sand."

I started the engine. It purred convincingly, no cutting out or vibrations. I taxied a few feet, still listening to the engine. Then I opened the throttle. Every time I hit an indentation or rise in the sand, I nearly choked.

When I ran out of solid ground, I bounced on some hummocks, and then eased off the sand. The plane staggered into the air. I'd made it — at least this far. Rather than shortcut across the water, I followed our pathway around the shoreline to the village — just in case.

Mishal and Lillian met me at the Point Hope airstrip when I landed. Roger returned later, in Enoch's boat, with our gear

Roger, Mishal, Lillian with an ATV in Pt. Hope, 1982

The rest of the day resembled what I'd imagined we'd be doing here; visiting with the Tooyak family and exploring the village. Every time Mishal was introduced as "Dora's daughter," eyes lit up and questions were asked, followed by extensive recounts of Tooyak family history. Roy Vincent, Mishal's step-grandfather, relished every minute reciting incidents in his life. All this took place over cups of strong tea, as well as walking tours around the old and new villages.

In the old village, which had nearly been washed out by fall storms, the sod houses remained semi-submerged in the tundra. Each house had a low, tunnel-like entrance, which led down into it.

Whale bones in Pt. Hope cemetery, 1982

Walking around the cemetery was a truly unusual sight. It was fenced with eight to ten-foot tall whale bones. With the lack of timber at this latitude, the Natives improvised with bow-head whale bones for a variety of needs. Their crafts reflected medium available in their environment, too. They were known for their skill at making whalebone masks, ivory carvings, and baleen baskets. I recognized that Mishal's artistic ability and adept fingers came from this heritage. Andrew and Irene directed us to Beatrice Tooyak's grave, Mishal's maternal grandmother, and the woman I had taken movies of my first trip to Point Hope.

In 1970, the traditional village had been re-located to the east. This area was more able to endure the fierce and devastating storms that swept out of the Chukchi Sea. The above-ground houses had been moved on runners; other houses were constructed by the Borough and by individuals. (Alaska is organized into boroughs, rather than counties.) The houses had attached mini-garages for three-wheelers and snow-machines, and many of them had televisions, which provided a window into a wider world for these isolated, top-of-the-world people. Just like Barrow, and even in the modern era of the 1980s, permafrost still determined the availability of services. There were no underground water or sewer systems. Indoor toilets were lined with plastic bags, which required daily disposal. Water was brought in from a lake, six miles to the east, and stored in individual tanks in home mini-garages. But there were no complaints and these chores were conducted matter-of-factly.

Roger still felt apprehensive about flying the 180 Cessna, so he, Andrew, and I had returned to the airstrip.

"You know, there could be structural damage," he said uneasily. "Maybe a cracked crankshaft or unseen propeller damage."

He examined the plane's prop and pulled up the cowling.

"I think we should file off the prop ends," he concluded.

Without conversation, we went to work. At intervals, we weighed the two sides of the prop, attempting the necessary balance. Once accomplished, Andrew pocketed the two ends.

"These are my souvenirs," said Andrew.

Better his memory than mine.

I'd planned to stay another day, but that morning while I was still in bed, I'd heard the wind whistle outside and noticed clouds gathering in the east. Yet, before leaving, we attended the services at the Episcopal Church, where the Tooyaks had a long history of leadership. Peel Tooyak, Mishal's grandfather, had been an interpreter, lay reader, and organist. Andrew and Irene were now elders. The congregation heartily sang all verses of "My Country 'tis of Thee," "How Great thou Art," and "Just as I Am" — first in English and then in Inupiaq.

We said farewells and a number of people followed us to the airstrip. Mishal had been abundantly immersed in the lifestyle of her people and now hugs, hand-shakes, moist eyes, warm wishes, and "come backs," concluded the poignant reunion. I'd inquired about a telephone to file a flight plan. There was one phone in the village, but repeated attempts to connect with Kotzebue failed. I watched to see if Lillian would have reservations about getting back in the plane. I couldn't catch her eye — and I didn't ask. We found our places and even though no one verbalized it, we were all praying as the plane left the gravel strip and climbed into the air.

"I don't think we should fly over the water," Roger advised, still anxious about the plane's airworthiness. I followed the coastline, not taking any short-cuts over open water. Rain seemed to be our constant companion and found us before we reached Kotzebue, and then followed us to Galena. Clouds piled high in our path homeward, so we settled down at Galena until I got a positive pilot report, after which we skirted the remains of the thunder-boomers and found Rainy Pass. This was definitely a trip when a co-pilot, not only a mechanic, came in handy. The pass was indeed rainy, and visibility was poor. I appreciated another set of pilot-eyes. Once again, we roller-coastered through the pass; all the while I hoped whatever might have been damaged in the plane engine would not shake loose.

As we drew closer to home, I crossed off the landmarks, relieved we'd made it this far. Like a celebrative receiving line, the oil platforms sparkled in the Cook Inlet, and lights blinked on ships docked at the Kenai Oil Refinery.

At 10 p.m., the sun had only begun to sag in the sky and I buzzed the homestead house, letting Ruby know we'd returned, and checking for dusk-grazing moose on the runway. Slipping down between the tall spruce, the travel-worn 180 settled onto Gaede Private airstrip, and like a faithful dog saw us to the hangar. We were home.

We climbed out, greeted by Ruby. Her flour-dusted apron showed beneath her unbuttoned jacket. Who knew what she was baking in the kitchen at this hour. Maybe rhubarb pie? We could only hope.

"How was it?" she asked, fastening her eyes on Lillian.

Lillian gave a half-hearted laugh. "Well, you were right."

"Come on in and tell me about it," Ruby said.

Roger, Mishal, and I unloaded the plane. We worked silently. I thought how Mishal, my youngest daughter, had shown her true nature, that of a dauntless, persevering Alaskan woman. Her survival instinct was innate. Roger, too, had been there when I needed him. Every mile of the way he'd sat alert as my co-pilot. His resourcefulness had gotten us out of a mighty bad pickle.

"Let's go see if there is rhubarb pie waiting for us," I said.

They nodded.

It was good to be home.

CHAPTER 26

FLIGHT BY FAITH

October 1982

THE ALASKA SUN only teased the 35° temperatures on this late October day. I carefully crammed lumber and other building supplies into my Cessna 180. I expected a routine flight across the Cook Inlet and through the mountains. My destination, 150 miles west, would lead me across the Inlet, through the Alaskan Mountain Range, and to Lake Clark, via the 60-mile Lake Clark Pass. The materials were for the house Ruby and I were building at Port Alsworth, near Paul and Irene Carlson's. After Irene retired from teaching at Newhalen School, on Lake Iliamna in1974, she and Paul built and ran North Country Lodge in Port Alsworth. I imagined Paul now, his red-and-black checked wool shirt and corduroy cap the color of sawdust. His hammer, help, and upbeat nature heightened my desire for these trips, and made the building task easier and more fun. I wouldn't turn down anything Irene might offer to feed me either.

Although the pass was frequently plagued by treacherous weather and although its floor was littered with numerous accidents, I felt confident about reaching my destination. After all, I'd flown these specific mountains for over 20 years and navigated through this pass 12 to15 times each year for the past several years. I figured I could fly it blindfolded.

FAA weather briefing indicated 2,000-foot broken clouds with occasional snow squalls at Port Alsworth. The pass was estimated closed to marginal, but because *marginal* was still flyable, I decided to check it out for myself. With this in mind, I swung up into my plane and took off from the grassy homestead airstrip.

After climbing to 5,000 feet over the inlet, I called Kenai radio.

"This is Cessna 9762 Gulf. Do you have any recent pilot reports through Lake Clark Pass?"

"Cessna 9762 Gulf, a Cessna 180 came through from Port Alsworth an hour ago and reported marginal conditions," crackled the reply.

A second voice entered the conversation.

"Cessna 9762 Gulf, call me on 122.9."

I changed frequencies.

"Doc, this is Jack. I just came through the pass an hour ago. It was marginal, but I made it okay. It should be improving. I'm on my way back. Do you want to fly it together?"

"Sounds good," I replied. I knew Jack Vantrease.

Like long fingers, gray clouds reached across the 8,000-foot high Alaska Range, ready to snatch planes that dared enter its inner sanctum. When I arrived at the entrance to the pass, the ceiling of 3,500 feet looked accessible and mocked my indecision.

"Doc, I'm about a mile behind you," Jack called over the radio.

"I'll go ahead and give it a try," I replied, accepting the challenge.

A few aspen, stubbornly refusing to relinquish their golden leaves to winter's grasp, speckled the mountain sides, and appeared out of place in the dark canyon. Flying several miles into the narrow east end, I turned left. Before me, a glacier crept across the center of the pass, accompanied by snow flurries. From this summit, a river flowed away, both to the east and west. The glacier had a bad reputation. In the winter, it often produced whiteout conditions which blinded pilots — their planes were later found crashed.

I dropped to 1,100 feet, gaining an uncomfortably closer view of the treacherous blue crevasses, yet able to fly on past the glacier.

Then, without warning, I was blinded! Hastily I made a180-degree turn. Enveloped in the snow flurries, I groped for direction. My turn-and-bank indicator told me I was not level, even though I felt level. Shocked, I realized I was flying vertically alongside the mountain, mistaking it for the floor of the Pass.

"Help me!" I gasped.

Reorienting myself, I flew back over the glacier where there was partial visibility. At once, I checked for carburetor icing and the insidious wing icing, which destroys the lift necessary to carry a plane through the air.

"Jack, it's bad in there," I called over the radio. "I've turned back."

"Okay Doc, I'll punch into it and give you my opinion. I've flown it so many times I can easily pick out landmarks."

Even with our rotating beacons, navigation lights, and landing lights, I did not see Jack pass by.

"I'm past the glacier and over the lake now," he called after a few minutes. "It's pretty bad, but I can pick out some trees on my left. I think we'll be all right. I'll guide you through. Are you still over the glacier?"

The valley was obscured and the possibility of losing my way was nearly certain. Could I trust Jack? It was true that conditions could improve. On the other hand, the pass threaded between 6,000- to 8,000-foot mountains. But then, I'd known Jack for over 20 years — he *was* a good pilot.

Deciding to trust him, I answered, "Yes, Jack, I'm still here."

"Fine. Head down the middle of the valley. Watch your compass and stay on that heading. Don't turn. No matter how wrong it feels."

Jack seemed to know what he was talking about.

"Keep your altimeter at 1,200 feet and airspeed at 150 mph."

With no sight of Jack's plane and only a voice to follow, I tracked down the middle of the canyon.

"Jack, I'm on the east edge of the lake. I can't see across."

"Doc, you're about a half mile behind me. You'll soon see a row of spruce trees to your left just after you cross the lake. Don't look ahead. Trust your instruments. Trust me."

I had almost no forward visibility, but out my side window I could distinguish the spruce trees lined up and pointing the way.

"Now you should see a line of trees on your right. Follow them. In another few seconds you'll see the edge of a hill on your right... Now make a slight correction with your compass heading to the right — only a few degrees."

I clung to Jack's words. My heart pounded. The landmarks didn't look familiar. Was this the right direction? Maybe I should turn to my left. How had I gotten myself into this?

Silently I shouted a prayer, "Oh, God, I made a mistake to enter this valley. Will you help me anyway?"

I was wedged in between the snow and the unknown. My hands, damp inside my warm winter gloves, clung to the stick. Carefully I maneuvered the plane. One wrong move and I'd be a statistic in this valley of death.

Elmer in cockpit

Time seemed suspended even though the plane fought through the choking blanket of snow at two-and a-half miles a minute.

Clearing my throat and wetting my lips, I radioed Jack. "What now? Are you picking up any ice?"

"Well, Doc, there is some, but not more than a quarter-inch on the wings and struts... You should be over the little pond on your right."

"Yes. I can see it about a quarter mile ahead now. What comes next?" I needed to hear his voice.

"In another minute you'll come to a branch in the river. If you follow what appears to be the river it will go to the right. Don't go that way — it's deceiving and leads to a dead-end. Remember the Beaver plane that cracked up in there last year? Instead, look at your compass and keep going straight, over the solid stand of trees. You may not see the river for half a minute, but don't worry... by the way, did you see the big bull moose below you?"

Don't worry? I wiped my forehead and strained my eyes for the river. I sure wasn't interested in moose right now. Finally, the river appeared.

"It's looking better ahead."

Jack's words kept me going. Sure enough, within a couple of minutes the snow let up and I could see ahead two miles.

"We made it!" I nearly yelled as I unclenched my jaws and relaxed my grip on the stick. I'd been holding my breath and now gulped for air.

The hours of anxiety dissolved into nine minutes of actual flying time — still long enough to allow for a fatal accident.

I sighed a quick prayer, "Thanks, God, for flying with me."

When we broke out of the canyon over Lake Clark, Jack indicated his own feelings, "Boy, Doc, that was worse than I've ever been through."

"I don't care to do a repeat on that. Thanks for talking me through, and I'm glad we both made it."

The rest of the trip to Port Alsworth was in sharp contrast to the previous flight, and as I slowly settled back in my seat, I recalled another close call with death.

In the early '70s, I'd volunteered to help MARC shuttle students from the Mission Covenant High School at Unalakleet (YOU-na-la-kleet) back to their village homes during the springtime. I was in my four-place Maule Rocket with a 220-horsepower fuel-injection engine. With some 1,200 hours flying time logged in Alaska and a couple 100 hours in this place, I felt at ease flying along the Alaska coast; even though flying out of Unalakleet, on the coast of the Norton Sound, was often a bit tricky without an instrument rating. Fog persistently afflicted this region.

I had three Native students with their abundance of gear. My first student, Betty, had enough energy to fill the entire plane. She hugged and chattered with the school teacher who had brought her to the airstrip; then she teased the two boys who were also on my passenger list. Before I even started the engine, I'd heard everything she planned to do when she got home.

She lived in the small Inupiat village of Selawik, 80 miles east of Kotzebue.

The FAA weather report between Norton Sound and Kotzebue Sound was marginal. In my favor, several pilots reported coming through without difficulty. I experienced the same favorable weather and while the students talked among themselves and pointed out the windows, I steadily worked toward my first destination. This village was difficult to locate during the winter because the snow-covered tundra and frozen lakes blended together, camouflaging the village cabins. Besides this sightseeing challenge, the village did not have an airfield.

Betty, sitting in the front seat beside me, was an eager-to-help co-pilot. She knew the landmarks, and found the village — not more than shadows on the opaque surface.

"There, Dr. Gaede. You can land there," she pointed to a nearby river, not expansive like the Yukon, but adequate.

This was breakup time and already there was water along the river edges, with occasional ice bridges to the shore from the decomposing ice in the center. It was just a matter of time before all the ice would let loose and churn down the river. I wasn't inclined to set down on this potentially hazardous natural airfield.

Even with her seatbelt on, Betty bounced in her seat, eager to get home.

"Sure, Dr. Gaede. It's alright. They do it all the time."

I wasn't sure who "they" were, but I saw a Piper Cherokee Six tied down on the edge of the ice. Flying over the area again, I detected a few wheel tracks on the center of the ice. With hesitation, I pulled full flaps and settled down — as lightly as possible. The ice held and I carefully braked to a stop.

By this time, a crowd of spectators lined the riverbank. I wasn't sure if it was just a welcoming committee for whoever might be stopping in or if they actually knew Betty and were there to greet her. Whatever the motivation, I wanted to believe that if we crushed through the ice they would come to our rescue.

Betty bumped hastily out the door and stood impatiently waiting for her duffel bag and box.

"Thank you, Dr. Gaede!" was the last I saw of the bright-eyed girl as she slogged through the slushy ice to the shore.

I took off for Kotzebue. My anxiety was lessened when the plane pulled above the rotten river ice.

The flight to Kotzebue was quiet, both inside and outside the cabin, and without delay the two boys were unloaded at the airport. Shyly they reached to shake my hand.

"Thank you, Dr. Gaede," they said in unison.

"That was a fine flight," added one.

Now I was on my way to Moses Point on Norton Sound. The flight to Moses Point was routine and after learning that no fuel was available, I closed my flight plan and continued to Nome for gas. One wing fuel gauge indicated plenty of fuel. I figured I'd have no trouble making it the additional 40 minutes.

As I crossed Golovin (GAWL-uh-vin) Bay, 30 miles toward Nome, the onshore weather deteriorated and clouds filled in the hills. Visibility appeared better over the water, so I swung out about a mile and cruised above the waves at 400 feet.

With no forewarning, the engine quit! Reactively, I turned the plane toward the shore. The bay was also in breakup condition, with numerous scattered ice flows, but none were larger than 50 feet in diameter — certainly not large enough for an emergency landing. Quickly I set up a glide. I was mystified. My fuel gauge showed one-fourth full. What could be wrong? Desperately I switched to the other tank, engaged the electric fuel pump, and pumped the throttle; then, in hope of draining the tanks, rocked the wings. No response.

Without power, the 155 mph cruise speed rapidly bled down to 80 mph, and the plane began losing altitude. Without power, I would be ditching about a half-mile from shore. The plane would sink in about three minutes and I'd last about two minutes in the frigid waters. There was no hope of surviving. My only prospect was an icy grave.

The airspeed declined to 60 mph and my altitude was less than 100 feet above the menacing waters.

"God, save me!" I prayed loudly.

The glide continued downward to less than 40 feet. In an attempt to prolong the flight, I pulled first notch of flaps. Touchdown was in less than five seconds.

Suddenly the engine surged to full power and the airplane began to climb. I was shocked and numb with joy. God had heard my prayer and saved my life! Then as I crossed the shoreline at 200 feet altitude, the engine again quit. This time, however, I was not terrified. Flat tundra with a cushion of snow spread beneath me, ready for an emergency landing. Just as I had done the first time, I rocked the wings to drain out any possible gas. The engine caught. I turned inland, located the small Golovin airstrip, and landed.

With shaking knees, I climbed out of the plane. Mr. Olson, the local air service operator, met me. Unnerved, I babbled my story to him. He listened, and then offered to gas my plane. I walked around in a stupor, glad to be alive and on dry land.

He finished fueling and climbed down the short ladder.

"You are very fortunate to have made it here," he said slowly, his eyes fixed on me in astonishment. "You used up five of your six unusable gallons of gas."

What? I couldn't comprehend what he was telling me. My fuel gauge had been in error. With God's divine help, I'd been able to slosh enough gas around in the wing tanks and down to the engine.

The sight of Lake Clark brought me back to the present. I started my descent and ended my ponderings about life and death. My relationship with God was one of faith, even though I couldn't see Him and even when I made wrong decisions. He was a constant companion who traveled through life's canyons and over life's rough waters.

CHAPTER 27

NO ORDINARY DAY

March 1984

JUST AN ORDINARY SATURDAY, I thought. As I'd done many times before, I planned to fly to Port Alsworth. Scattered clouds rested benignly against the clear sky and brisk, invigorating air greeted me as I strode out the front door in my typical flying gear of flannel shirt, blue jeans, and insulated coveralls. My heavy work boots pressed into the damp gravel circle drive, certainly a welcome sound after the winter-long crunch of snow.

Instead of going to my hangar, I proceeded toward the road in front of the house, a road that would later acquire the legal descriptor of *Gaede Street*. It was too early in the spring for chickens to be in the chicken coop nearby. Between the wide front lawn and the road grew a miniature forest of spruce and birch, which shielded the house from traffic on the road. Rising above the grove, the orange windsock waved gaily. Mark's pumpkin-colored J-3 sat by itself to the left. Across the road was the A-frame where Mishal lived; the steep green-shingled roof was trimmed in orange. My children laughed at all the orange around the homestead, but I'd gotten a good deal on a large quantity at Penn's Hardware. Besides, who would see the orange back door of the main house?

During this breakup season, the homestead airstrip, which was dirt and not gravel, turned into a lake with an unstable boggy bottom; thus, Roger, Mark, and I used the front road for a temporary landing strip. This morning, a total of four planes, including my newly painted white-and-blue Cessna 180, stood at attention like sentries guarding the entrance to our homestead. I'd owned the 180 for nearly three years now, but still found it difficult to land. It bounced easily when it touched the ground, which was an aggravating response.

I tossed an extra parka into the back of the plane. After buckling in, I started the engine, checked for cars on the road, and slowly taxied to the south end of the road. I passed my Super Cub, a patriotic aircraft with red and blue strips on white; followed by Roger's freshly painted dark blue with yellow and white stripes Cessna 172, then spun the plane around before I reached the Gas Well Road and Jones Stub intersection. The neighboring homesteaders had learned to be cautious when entering this double-duty road.

Except for some springtime gusty winds, it was a perfect day. Adding "watch for cars" to my preflight list, I checked the magnetos and prepared to take off. The plane, heavily loaded with lumber, accelerated down the "runway" and lifted off. The ground rapidly dropped away.

I headed to Kenai and then across Cook Inlet toward Lake Clark Pass. I anticipated getting into Port Alsworth without any weather problems, but, true to its reputation, shortly after entering the pass, a snowstorm enveloped the plane. I did a 180-degree turn, headed back to blue skies, and called Kenai radio with a report of the actual conditions in the pass.

"I'll fly around for half an hour before trying again," I told the weather briefer.

I figured my chances were 50-50 I'd have clear sailing soon.

As I flew around, I thought of all the planes with which I'd cut ruts in the Alaskan sky. Prior to this one, I'd had the powerful Maule Rocket, which I nearly ditched into the Norton Sound, and which two years previously, in 1967, I had cracked up on the Kenai Peninsula.

I'd been coming back from a staff meeting at the hospital in Seward and lost power at 400 hundred feet. No sudden surges of power pulled me out of that one. I pancaked into muskeg and bounced into spruce trees. Shoulder harnesses weren't worn then and I'd buried my face into the instrument panel. My passenger, Lee Bowman, the Soldotna pharmacist, sustained a broken ankle, and I had a compression fracture of the upper lumbar vertebra. Although I was in good hands with the well-qualified clinic nurses, Dr. Isaak hastily returned from Seward, in his plane, and sutured my face back together. It took over 50 stitches to pull together my forehead, chin, and mouth area. Five teeth were knocked out. Since I couldn't shave over this new face-lift, I grew a goatee, mustache, and long sideburns.

That was a sobering experience. Just like a bucked rider has to get back on a horse, I knew I had to get back into the air. After a month, I went up with a friend in his airplane. Obviously, I was back riding in the big skies now.

As I had predicted, the snowstorm dissipated. Within 20 minutes, my plane zigzagged through the pass and I landed on Babe Alsworth's lake-side airstrip. Watching my time, I worked until around 3 p.m., and then headed back to the Gaede Eighty homestead. The plane, like a homing pigeon, could probably have flown home alone; the flight was uneventful.

At 5,000 feet over the inlet, I started to pick up light turbulence, but nothing to be alarmed about. I buzzed the house at 4 p.m. I knew Ruby would recognize the sound of my plane, but I doubted her usual concern about my return would subside until she saw me taxi to my tie-downs. Just the day before, Mark had walked in white-faced and silent after making a road landing in the Cessna 180.

I flew over the house and noted the windsock, which showed 10 to 15 knots. I decided to make one more go-around before my final approach. I had never landed the 180 from the south. Passing over the house, I saw Ruby's father, 88-year-old Solomon Leppke, start toward the circle drive, walking slowly with his diamond willow cane. The old gentlemen loved Alaska and had been coming here from Kansas since he was 60 years old. The tough farmer never backed away from hard work and had put his sweat into nearly every project on the homestead. He most likely wanted to see me land. I made my final approach to the angled homestead road. Coming in above the 60-foot trees and easily clearing the power lines over the road, I chopped the engine, expecting to glide about 100 feet.

The plane dropped like a rock! There was no air to hold me! My left wing clipped the tops of the 15-foot spruce trees flanking the road. I made an off-balance touchdown, but the confrontation of my leading edge with the one-inch-diameter trees had pulled the aircraft to the left. I found myself heading directly toward Roger's freshly painted 172! Terrified, I slammed the throttle forward in a desperate attempt to reestablish airspeed. The Cessna 180 had a lag and took precious seconds to respond.

From years of flying, my mind made instant reflex decisions. I evaluated the parameters of heavy overhead wires, trees, and airplanes — all around me. Frantically I maneuvered the yoke at 45 mph, barely missing the guy wires of the light pole. Miraculously, I leap-frogged both Roger's 172 and my Super Cub. Time moved in slow motion as I headed toward our front yard. When I'd cleared the Cub, the 180 gained an altitude of about 20 feet. Then, to my alarm, I severed our telephone line with the propeller. The aircraft was really floundering now.

After surviving the Maule wreck, I knew it was possible to survive a controlled crash at low altitude with slow speed. But, with my airspeed near stalling, and with full power, the prop torque kept me from gaining control. I was living a recurring nightmare, familiar to all bush pilots — the panicked feeling of being boxed in. Suddenly my left wing hit a birch, tearing off five feet of wingtip. My ailerons went with it. I was now above the front lawn.

Later my family told me what they experienced in their ringside seats. Ruby, hearing the sound of a plane aborting landing, rushed from the kitchen to the picture window in our living room. The window served as an enormous television screen portraying a horrifying plot — a drama she could not turn off. My plane hurtled across our lawn. Never, had she imagined I'd crash before her very eyes.

"Mom!" she'd cried out at her mother, Bertha, who was indoors with her, "Elmer is crashing!"

Solomon Leppke stood on the edge of the drive, 45 feet from where I'd again touched down. The odd free-form flight of my disoriented bird coming at him stopped him in his tracks.

Roger was at his house washing his truck. He heard the scream of my plane in trouble, jumped into his pickup, and raced to our front road.

Meanwhile, with only ten feet of altitude, I still had hope that I could get around the small grove of trees and return to the road. I had to disengage myself from this bizarre steeplechase! I calculated that if I could pull to the right, I could make it out the driveway, dodge Mark's J-3, and bring the plane to a stop — without crashing. This could have worked, but my airspeed had dropped to a stall, and my damaged wingtip and aileron made control impossible. Then I saw the massive telephone pole loom in the way of my intended escape route. My choices were gone. I opted for the trees and chopped the power.

The plane tore down our four-foot garden fence, hit one tree, smashed into four others, and came to rest near the chicken coop. The sounds of my bruised black and blue aircraft ended in deathly silence.

Ruby ran out of the house. Her nightmare, as a bush pilot's wife, was coming true: I'd die in an airplane crash. As my daughter, Ruth, said later, "We'd always expected Dad to have another plane wreck, but we thought it would be across the inlet, not in our front yard."

When Ruby neared the coop, where the plane now roosted, she saw that a wheel was ripped off, both cabin doors were torn away, and the windshield was popped loose. Uninvited spruce branches filled the cabin. Unaware of the blood dripping from my head, I climbed out of the plane, stunned and amazed that I was still alive.

"I'm okay."

Cessna 180 crashed by chicken coop, March 1984

"You're not okay," she protested. "You're bleeding. You cut your head."

"I'm all right!" I loudly claimed.

In hysterical relief we argued.

Mishal, who had been in Soldotna and had missed the drama, drove down the road — oblivious of the turmoil that had preceded her. She parked at the A-frame and ambled up the driveway. Unaware of the plane lodged in the chicken pen, yet sensing something out of the ordinary, she casually asked, "What happened?"

In babbled sentences, we all tried to explain.

At the same time, Roger abandoned his efforts to find my plane and in bewilderment drove up the drive, right past the well-camouflaged 180, which hid in the woods like a wounded animal. My family clustered around me.

Conscious now of the warm blood running down the side of my face, and after Mishal's examination of both the three-inch and four-inch gashes in my head, I allowed myself to be taken to the emergency room. At least a shoulder harness had saved my face this time. Thirty stitches later, Mishal drove me home. I saw the litter-strewn pathway of my kamikaze landing.

Before retiring to the house for the evening, I picked my way back to my plane. Poking my head inside the door-less cabin I saw several strands of gray hair caught on a protruding screw, on the opposite of the cabin from where I'd been sitting. It fluttered gently in the air. It had been a close call.

For months, I would ask other pilots what I should have done. They all agreed that there were no other options and no "if onlys." They also agreed that gray hair comes with the territory. With deep gratitude, I thanked God for being in control of an out-of-control situation. My number was not up.

No, March 3, 1984, was no ordinary day.

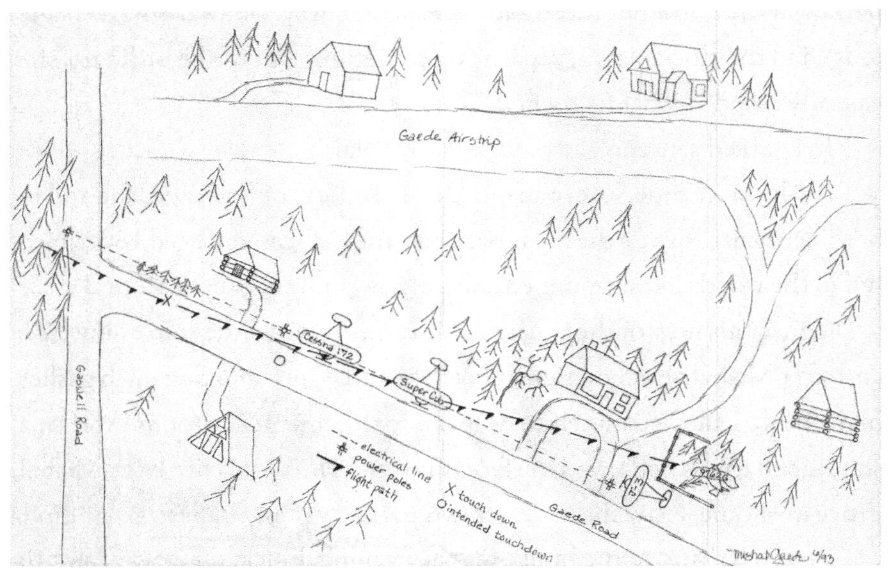

Elmer's flight path off Gaede Street and across the Gaede's front yard, March 1984

(Courtesy of Mishal Gaede)

The Story of the Cessna 180

The Cessna 180 was originally Don Sheldon's. (See chapter 6, "Out to Get a Bear Rug" when Doc met him at Talkeetna.) Sheldon had owned Cessna 9762G and 9763G. Both were bought by the Lofstedt family of Kenai Air. Elmer Gaede bought 9762G and the Lofstedts kept 9763G, which Vern Lofstedt owned until his death in 2011; after which it went to Trevor L. Lofstedt in Scottsdale, Arizona. Sheldon had initially bought both airplanes bare, and sold them that way. Roger painted both planes. The 9763G number was changed to N180KA for Kenai Air. After Vern Lofstedt acquired it, Vern asked Roger to paint it in the Dr. Gaede 180 (62G) paint scheme in memory of Doc. He had lettering put on the cowling "In Memory of Dr. E. Gaede."

Cheechako News

August 4, 1967

Forced Landing Injures Soldotna Doctor, Drugist

A Soldotna doctor was the pilot in a four-place Maule Rocket, which crash landed Wednesday morning about four miles off the Sterling Highway out of Soldotna, injuring two of the three persons aboard.

Dr. Elmer Gaede, pilot, and Lee Bowman, local pharmacist, both received multiple lacerations of the head and face and nearly identical injuries to the spine. Dave Parks, who came to Alaska about a month ago from Michigan, was seated in the back seat and was uninjured.

Dr. Gaede was returning from Seward where he and Dr. Paul Isaak, his associate in the clinic at Soldotna, had been engaged in surgery at the Seward Hospital.

Engine failure is believed to have been the cause of the forced landing which occurred at about 10:30 a.m. between the homes of Dr. Calvin Fair and "Chuck" Foster on Forest Lane. The plane was damaged extensively.

Joe Norris and Dave Thomas, employees of Soldotna Supply, were unloading lumber at the Fair home when the plane flew over at about 500 feet altitude. Shortly thereafter, one of the Foster boys ran to notify them that the aircraft had crashed nearby. The Soldotna ambulance was called and the men transported to the clinic where Dr. Robert Beckman and the X-ray technician took charge.

Dr. Isaak was summoned and arrived shortly after Bowman had been flown to a hospital in Anchorage by Troy Hodges, who operates a local flying service. In his examination at the clinic, Bowman was found to have a compression fracture of the spine and possible other injuries in addition to lacerations.

After receiving treatment, Dr. Gaede was taken to his home near Soldotna where he is expected to have about a month's time to catch up on his neglected reading before returning to his practice. His injuries, in addition to lacerations of the head and face, include contusions in his left shoulder, left arm and left leg and two fractured lumbar vertebrae.

CHAPTER 28

THE VALLEY OF 10,000 SMOKES REVISITED

August 1986

KNOWING THE CAPRICIOUSNESS of Alaska weather, I'd booked off consecutive Fridays in hopes of finding one long weekend to fly to the Valley of 10,000 Smokes — located across Cook Inlet and to the southwest near King Salmon. So far, the only thing the crossed-off Fridays indicated was a long spell of objectionable flying conditions around King Salmon.

At this stage of my medical career, I was no longer delivering babies, so an unexpected phone call wouldn't complicate my plans. In fact, the last one I'd delivered had been Erin O'Brien, born to Timothy and Roberta O'Brien, on May 2, 1985. Cutting back was a difficult decision since my passion for medicine never waned. I gained pleasure in bringing new lives into the world and seeing tears of joy on a parent's face. And, I continued to be intrigued by diagnosis and I'd learned that listening to a patient's symptoms lent as many, if not more, clues than a laboratory report — which then confirmed, or confused what I'd heard.

Even after I'd retire in 1987, I would continue treating the Native population at a clinic held in the dispensary at Wildwood Army Base in Kenai.

My concern about practicing medicine, however, had to do with socialized medicine and insurance companies controlling my decisions about tests to order, procedures to follow, and treatment. This concern had increased. Not to mention rocketing malpractice insurance. All in all, I didn't mind some respite; after all, I'd been on-call most of 27 years of my 31-year medical career.

In the spring, I'd remarked to Leonard Olson, "July would be a great month to go to the Valley of 10,000 Smokes. The sockeye salmon will be running in the Brooks River, and we'll have plenty of daylight hours for flying there and back."

"Whenever you say the word, I'll be ready," replied Leonard, nodding his balding head in enthusiasm.

I'd chosen Leonard, a retired school principal and long-time acquaintance, to be my partner for this trip. He was a pilot, thus a bona fide co-pilot if I got in a pinch; plus, we both had interests in geology and ecology. With his flexible schedule, he could accompany me at nearly a moment's notice.

The Valley of 10,000 Smokes is no typical wilderness area, nor is it a common arctic land-form. Situated within the Katmai National Park and Preserve, the Valley of 10,000 Smokes is the site of one of the most violent volcanic eruptions in history. In June 1912, Novarupta ("newly erupted") Volcano exploded with a blast of hot wind and gas. Spewing immense quantities of glowing pumice and ash over the terrain, the volcano buried more than 40 square miles of dense green valley. Some of these lava deposits were in depths of 700 feet. Several years after this cataclysmic event, Dr. Robert Griggs, an explorer from the National Geographical Society, named the ash-filled valley with countless steam fumaroles "The Valley of 10,000 Smokes."

As I looked forward to exploring this wonder of the Alaskan world, July came and went; and the cloudy Friday weather pattern of Instrument Flight Rules (IFR) conditions, high winds, and severe turbulence camped out in the King Salmon area. In the interim, my desire to fly to the Valley of 10,000 Smokes grew into an obsession. Restlessly, I crossed off the first Friday in August. Daily, I watched the weather. Even though I knew better than to push into foul weather, and even though there were plenty of fair-weather places to investigate, I was compelled to re-explore this area. Finally, on August 15, my routine Friday noon call to FAA informed me that King Salmon had clear skies.

Immediately, I picked up the phone and dialed Leonard.

"All systems are go! I'll meet you out at Longmere Lake."

I traveled light. It didn't take long to grab necessities.

Just as I'd crawled back into a plane after my first crackup with the Maule, I was soon in the air after my unexpected flight into the chicken coop. This time, I was in a white with blue trim Super Cub, which, depending on the time of year, had floats, wheels, or skis. At the moment, it was on floats.

My first exposure to the Valley of 10,000 Smokes had been 30 years prior. That time my origination point was Lake Hood in Anchorage. Paul Carlson, who had since grown into a lifelong friend, had shared that quest in the J-3.

"How would you like to go with me to see the fabulous Valley of 10,000 Smokes?" I'd asked him one night as I poured over flight charts.

"Sure, lead the way," he'd said, chuckling.

We acted like two boys about to go over a country hill to a new fishing pond. In those early years of my Alaska life, I was only occasionally scared, often caught off guard, young with optimism, and inoculated with the belief that I was at least somewhat invincible. At that time, Alaska hadn't tested me with her capricious ways and the forces of her nature; consequently, I was a bold pilot. Odds always catch up, and seldom do brash boldness and oldness go together in the far north wilderness. I was naïve.

All that aside, the clear blue sky on that late summer day buoyed us forward. Truthfully, it was our second attempt to take off that day. I'd had the good sense to carry along extra fuel and we'd crammed four, five-gallon cans of gas into the minimal storage area behind the back seat. When the control tower had flashed the green light, I'd given full power to the 75 hp aircraft. Try as I might, I couldn't gain the necessary acceleration and the floats wouldn't lift onto the step. Finally, resigning myself to the suction-tight mirror surface, the heat of 70° temperatures, and, the overload of gasoline, I returned to the dock and removed two of the cans. That made the difference and this time I coaxed the floats off the surface and into the air.

In those early days of flying, venturing beyond the parameters of Anchorage was a pretty big deal; notwithstanding, without navigation or weather problems, we crossed Cook Inlet, through Lake Clark Pass, and to Iliamna Lake. Here we refueled and then continued toward King Salmon. We landed on the aquamarine Naknek Lake at Brooks Camp.

A friendly Fish and Game ranger met us at the shore and directed us to a good tie-down spot. He filled us in on bits of information and whatnot as we settled in, and as soon as we had the J-3 settled in, we pulled out our fishing gear. The rainbow and grayling trout were hitting the flies. Our fishing dreams were becoming reality.

"Paul. This is a fishing paradise!"

The next morning, we flew the short distance east to the Valley of 10,000 Smokes. The Katmai volcano and craters were astonishing and we looked in disbelief at the scorched terrain. At the west end, a glacier melted into the valley, which was a bare desert of lava. A dozen active fumaroles in the upper valley spurted steam. In slow flight, we zigzagged until our curiosity was met and then returned to camp.

After topping one fuel tank with gas from the two cans we'd brought along, we climbed back into the air. We figured our return home would be a piece of cake and we settled in to enjoy the "Alaska Calendar" scenery.

We were mistaken.

Between King Salmon and Lake Iliamna, the winds picked up with a vengeance, tossing the frail J-3 about like a salmon in a grizzly's jaws. My camera whizzed around the cabin, as did the fishing gear; all those hard-edged items became a danger to both of us. Paul attempted to anchor himself to the seat and at the same time catch and secure the identifiable flying objects.

I was busy at the few controls the J-3 offered. By calculating our ground speed, I realized we were having 50-knot winds at 2,000 feet. With this kind of opposition, we would be pinched on gas to reach a refueling place. I searched the lakes beneath us for a safe harbor until the winds subsided. I made a decision out of inexperience, and chose a lake for its length, not for its smooth surface. From my perspective, the waves did not appear dangerous. Innocently, I set the airplane down on the water.

The waves violently shook the plane and large white caps pounded at the floats. I idled the engine and then bent around to discuss the situation with Paul. It was then, with my head turned, that I recognized the impending danger. Out the back side window I saw the wind was shoving us toward the beach. This was not a bare-footing beach, but one stacked with enormous boulders. Within moments our plane would be pounded to pieces by these inhospitable giants.

"We've got to get out of here or we'll be crushed!" I yelled.

I jammed the throttle forward and pulled back slightly on the stick. The prop wash and the wave spray completely covered the windshield.

The plane floundered in the troughs of rough water. Blindly, I tried to feel my way back onto the "step." The waves crested and broke as I continued to ease the stick back. I hoped to bounce clear of the ensuing waves and increase airspeed, while at the same time prevent the plane from stalling. Then, following one gigantic teeth-jarring bounce the windshield cleared and we were airborne.

Tight-jawed, I'd urged the J-3 higher into the air. Only after I felt more secure did I once again turn back to Paul.

"We'll have to fight out this storm up here and hope to make it to Iliamna for gas."

That story had a happy ending and now three decades later, it was a quiet day and the Cub rocked easily on the lake in front of a friend's house. Leonard was waiting, his denim jacket tied around his waist. The toes of his hiking boots were damp from walking around the water's edge. We quickly assembled our gear. Just as on my first trip, we crossed Cook Inlet; unlike then, it was now dotted with oil well platforms. We flew into the notorious Lake Clark Pass, which this time was on its best behavior and smooth, dry air carried us through.

Before long, we splashed down on Lake Clark where I had aviation fuel stored at our house there. Then we continued over Iliamna Lake and straight toward Brooks Camp in the Katmai National Park on Naknek Lake.

As we neared our destination, the late afternoon sun reflected off the windswept sandy lava floor in sparkling hues of yellow, gold, and light brown. Within moments, the floats skimmed over the quiet water before settling down and gliding toward the smooth beach.

"A perfect day and a perfect landing," I commented to Leonard.

"Yes, and look how close we are to the campsites — no more than a 100-feet away."

The park ranger, a smiling young woman with a round flat-brimmed hat, walked down the beach and filled us in on the pleasures and problems of camping there.

"There are about 28 campsites, which you can see are well-cleared and level." She pointed a short distance away. "There are three shelters for eating and drying out wet clothing. And over there are two food caches. This summer we had 24 brown bears enjoying our river, but now since the main run of salmon have already spawned, only three remain."

The caches, with metal ladders, stood on skinny legs about ten feet above the ground, making it impossible for a bear to have a midnight snack.

After pitching our tent, we walked to the Visitor Center to attend a slide and film presentation of the Katmai National Park.

"Earthquakes had warned the villagers along the Katmai coast of the impending danger and many had fled their homes," the narrator told us. "But even with these precautions, many were caught in ashy darkness for three days and nights while the volcano hurled out the hot debris. According to the U.S. Geological Survey, this was the largest eruption in the world in the 20th century. At Kodiak, 100 miles southeast of the eruption, the 60 hours of darkness was so complete that a lantern held at arm's length could scarcely be seen."

Before returning to our campsite, another pilot found me and advised me to move my plane from the bay into a sheltered spot in the Brooks River.

"Sudden winds come up frequently, which can easily damage a plane."

The bright moon, which was climbing up among the stars in the clear sky, seemed to deny such a possibility; nevertheless we moved the plane, then called it a day and slept soundly until a splashing noise awakened us. I reached for my glasses and poked my head out of the tent.

"Looks like it's breakfast time — at least for that bear," I said, watching the bear slap a salmon with his enormous paw and then rip it in half with his massive jaws.

"I doubt that we're invited to his salmon buffet," replied Leonard.

We broke camp, carried our belongings to the plane, and headed to the main lodge for our tour. While we waited, we looked over the rest of the facilities. There were 16 modern guest cabins, with eating facilities in the lodge. In a brochure, we read about the numerous fishing and sightseeing trips available for the nearly 4,000 people who visited here over the past year. Several small planes flew in daily from King Salmon. Obviously, since Paul and I had been here years ago, this place had grown into quite a business.

Two eight-passenger, four-wheel-drive vans pulled up in front of the lodge. Each was equipped with a front-end winch and a cable hook in the rear.

One of the drivers noticed our interest and explained. "We always drive together in case of vehicle trouble when we cross the rivers."

Enthusiastically we climbed into the vans, which started down a very narrow one-way dirt road. The path twisted through heavily timbered woods where most of the trees were 40-foot white spruce. Spongy caribou moss and dark blue-black crow berries carpeted the forest floor. Crowding together on this cushion were tangled alders, dense willows, and aspen. On several occasions, the buses stopped and allowed us to hike to scenic lookouts or dashing waterfalls.

By noon, we completed the 23-mile drive and arrived at a Park Service cabin, sitting on top of Overlook Mountain on the edge of the spectacular Valley of 10,000 Smokes. The sun shone brightly over the expansive 14-mile long, 5-mile wide valley.

"We couldn't have asked for a better dining view," I remarked to Leonard as we pulled ham sandwiches out of our sack lunches.

"This is the third day like this during this summer," the driver-guide announced.

I nodded my head, visualizing the row of crossed off Fridays on my calendar.

"It is thought that, in the 1912 eruption, when four cubic miles of molten rock were ejected from the Novarupta vent, it drained a connecting underground 'plumbing' system to the Katmai magma reservoir so that the Katmai dome collapsed," our guide started to lecture. "This formed the Katmai Caldera and Lake, which have been slowly filling with water at the rate of about 6 to 35 feet per year. Some people believe that the lake will be filled by 2020." (In 2023, this still had not happened.)

We finished our lunches, took pictures, and continued listening to the guide.

"Even though upon the first exploration the valley was called 'The Valley of 10,000 Smokes,' it was later recognized as really being steam instead of smoke. Someone has speculated that this misnomer probably comes from the difficulty some of the Native villagers had in saying 'steam' as compared to the ease of saying 'smoke.'

I glanced over at Leonard, who listened with rapt attention.

"There are about 20 spots where the fumaroles make wet spots in the sand. Originally, the fumaroles were up to about 900° Celsius," the guide kept his lecture going. "Cooled by the rain and snow, the heat has dissipated."

"We can vouch for the rain," said one of two weary and bedraggled backpackers, who had just climbed up from the valley floor. "We spent over half our time in our tent because of either rain or high winds — there is no protection out there."

We left them as they were hanging their damp shirts over tree limbs, and hiked one-and-a-half miles down to the valley floor. Orange, yellow, and blue altered pumice marked the fumaroles, which had at one time crisscrossed the valley floor. The abrasive combination of tumbling rocks and pumice, water and high winds had eroded a river and canyons. The guide referred to these canyons, which were a couple hundred feet wide, as the "mini Grand Canyon." A mile from the base of the sandblasted valley, a small amount of green vegetation was starting to heal the lava burns and brighten the lunar-looking terrain. Horsetail and deep pink fireweed sprouted in nooks and crannies.

Leonard Olson at the Valley of 10,000 Smokes, August 1986

The tour concluded, and by early evening we were back in the sky.

"I never would have believed that Alaska had such a place," reflected Leonard as we circled an island-studded lake before heading home.

"The first time I came here it was a real surprise to me, too," I said.

The often-turbulent skies were peaceful, with only a minor crosswind; and for a long time we flew in silence, enjoying the reverie and majesty of a perfect Alaska day, complete with snow-capped mountains, autumn-colored swamp-edged lakes, and views for miles in all directions. Lake Clark Pass graciously let us back out over Cook Inlet into a peaceful sky of golden-edged pink clouds. I felt gratified I'd had the opportunity to revisit this unique area, to relax and enjoy Alaska — without any great drama to experience or to turn into a story.

I'd come north as a simple farm boy, urged on by a nurse to try out Alaska. I'd caught Frontier Fever and never gotten over it. As an adventure-seeker, I had hopes and dreams of exploring the remote Alaskan territory, and since the early 50s, and little by little, with airplane after airplane, I'd carved pathways over and into this great land. I loved her wild beauty and ruggedness. Her surprises of weather, geography, tundra and forests, wildlife, and people kept me guessing, yearning, curious, and at the same time, content. Even though I'd become familiar with much of the vast land, I was not bored. There was always one more mountain to climb, rocky edge to look over, volcano to check out, or landmark to investigate.

If anyone asked me, I'd still have to say that the prescription for adventure was Alaska.

EPILOGUE

DOC'S LOVE FOR ALASKA never ended. The day before he died, he flew across Cook Inlet to Port Alsworth to the house he was building there. The next day, Sunday, he taught an adult Sunday School class at Soldotna Bible Chapel, then drove home, took a nap, and died of a heart attack.

His family always expected him to die in an airplane crash, yet his death was undramatic, and at the age of 69, on October 6, 1991, he took his final flight to his eternal home.

Although he retired from his full-time practice in 1987, he remained active as a medical director of the Heritage Place nursing home in Soldotna, and offered medical assistance at the Dena'ina Health Center in Kenai. He also donated finances and labor to the construction of Cook Inlet Academy, a Christian school two miles from their house, and served on the school board.

The spring before his death, he had gone to Haiti on a medical mission trip and traveled to villages to provide medical treatment.

On October 7, the day after his death, he was scheduled to give a lecture on skin cancer at the Soldotna Senior Center.

At the time of Elmer's death, Ruby was in Kansas with her mother, Bertha Leppke, who was dying from a stroke. Ruby died of cancer on January 25, 1995, at the age of 71.

Elmer and Ruby were charter members of Soldotna Bible Chapel and contributed to the building program with finances, physical labor, and leadership. Ruby always had a heart for missionary women and started the Missions program. Given her experiences along the Yukon River, she understood the needs of families and led the group in putting together packages of encouragement, such as Christmas boxes with food and other items that were difficult to obtain in the village. These boxes were flown to the villages by Missionary Aviation Repair Center (MARC). MARC families were invited to live in the Gaede's cabin on the Gaede-Eighty homestead, which they did until after Ruby's death.

Elmer and Ruby had come to Alaska to serve people, and they did so, tirelessly and enthusiastically, until they died. Truly, they met their Maker who said, "Well done, my good and faithful servants." (Matthew 24:23)

The Gaede-Eighty homestead was passed on to their children where Mark (Patti) and Ruth (Roger) live. Naomi, whose primary residence is in Colorado, rebuilt the cabin, which burned in 2005, and visits "home" as often as possible. Mishal gravitated to Fairbanks where she is an advocate for her Native people as a Court Facilitator for Tanana Chiefs Conference in Tribal Government Services. The Gaede children cherish deeply their family's Alaska heritage and the Gaede homestead.

RESOURCES AND FURTHER READING

Alaska Almanac. Portland, OR: Alaska Northwest Books, 2013.

Alaska's Kenai Peninsula: The Road We've Traveled. Kenai Peninsula Historical Society: 2002.

Once Upon the Kenai: Stories from the People. Kenai Historical Society: 1985.

Snapshots at Statehood: A Focus on Communities that Became the Kenai Peninsula Borough. Kenai Peninsula Historical Society: 2009.

A Century of Faith: Centennial Commemorative, Episcopal Dioceses of Alaska 1895 — 1995. Fairbanks, AK: Centennial Press, 1995.

Anderson, James and Jim Rearden, *Arctic Bush Pilot*. Washington: Epicenter Press, 2000.

Billington, Keith. *House Calls by Dogsled: Six Years in an Arctic Medical Outpost*. Maderia Park, B.., Canada: Harbour Publishing, 2008.

Boylan, Janet (compiled by). *The Day Trees Bent to the Ground: Stories from the '64 Earthquake*. Anchorage, Alaska: 2004.

Bower, Charles D. *Fifty Years Below Zero: A Lifetime of Adventure in the Far North*. Fairbanks, AK: University of Alaska Press, 2004.

Brown, Altona. *Altona Brown: A Biography — Ruby*. Fairbanks, AK: Spirit Mountain Press, 1983.

Bruder, Jerry. *Heroes of the Horizon: Flying Adventures of Alaska's Legendary Bush Pilots*. Anchorage/Seattle: Alaska Northwest Books, 1991.

Campbell, John Martin, ed. *In a Hungry Country: Essays by Simon Paneak*. Fairbanks: University of Alaska Press, 2004.

Campbell, John Martin. *North Alaska Chronicle*. Sane Fe: Museum of New Mexico Press, 1998.

Cline, Michael S. *Tannick School: The Impact of Education on the Eskimos of Anaktuvuk Pass*, Anchorage, Alaska: Alaska Methodist Press, 1975.

Cohen, Stan. *The Great Alaska Earthquake*. Missoula, MT: Pictorial Histories Publishing Co., 2000.

Dart, Chuck and Gladys. *A Biography: Chuck and Gladys Dart — Manley Hot Springs*. Fairbanks: Spirit Mountain Press, 1982.

Fejes, Claire. *People of the Noatak*. Volcano, CA: Volcano Press, 1964.

Fejes, Claire, *Villages: Athabaskan Indian Life Along the Yukon River*. New York: Random House, 1981.

Fortuine, Robert. *Alaska Native Medical Center: A History*, 1953-1983. Anchorage: Alaska Native Health Center, 1986.

Fortuine, Robert. *Chills and Fevers*. Fairbanks, AK: University of Alaska Press, 1989.

Fortuine, Robert. *Must we All Die?* Fairbanks, AK: University of Alaska Press, 2005.

Gaede-Penner, Naomi, ed. *Honoring our Sacred Healing Place: Tanana, Alaska Development, History, Community, & Cultural Significance of the Tanana Hospital Complex*. Anchorage, AK: Alaska Area Native Health Service, 2008.

Gaede-Penner, Naomi, *'A' is for Alaska: Teacher to the Territory*. Mustang, OK: Tate Publishing, 2011.

Gaede-Penner, Naomi, *'A' is for Anaktuvuk: Teacher to the Nunamiut Eskimos*. Mustang, OK: Tate Publishing, 2016.

Gaede- Penner, Naomi, *From Kansas Wheat Fields to Alaska Tundra: a Mennonite Family Finds Home*. Mustang, OK: Tate Publishing, 2015.

Gaede-Penner, Naomi, *The Bush Doctor's Wife*. Denver, CO: Prescription for Adventure: 2021.

Griener, James. *Wager with the Wind: The Don Sheldon Story*. Chicago/ New York: Rand McNally & Company,1974.

Griffin, Joy (compiled by). *Where were you? Alaska 64 Earthquake.* Homer, AK: Wizard Works, 1996.

Harkey, Ira. *Noel Wien: Alaska Pioneer Bush Pilot*. Fairbanks, AK: University of Alaska Press, 1974.

Herron, Edward A. *Wings over Alaska: The Story of Ben Eielson. Pocket Books, 1968.*

Huntington, Sidney and Jim Rearden, *Shadow on the Yukon*. Portland, OR: Alaska Northwest Books, 1993.

Jordon, Nancy. *Frontier Physician: The Life and Legacy of Dr. C. Earl Albrecht*. Fairbanks/Seattle: Epicenter Press, 1996.

Keith, Sam and Richard Proenneke. *One Man's Wilderness*. Anchorage, AK: Alaska Northwest Books, 1999.

Madison, Curt and Yvonne Yarber, ed., *Josephine Roberts: Tanana.* Fairbanks, AK: Spirit Mountain Press, 1983.

Mathews, Sandra K., *Between Breathes: A Teacher in the Bush.* Albuquerque: University of New Mexico Press, 2006.

Morvius, Phyllis Demuth ed. *When the Geese Come: The Journals of a Moravian Missionary, Ella Ervin Romig, 1898 — 1905*. Fairbanks: University of Alaska Press, 1997.

Pedersen, Walt and Elsa. *A Larger History of the Kenai Peninsula*. Chicago, IL: Adams Press, 1983.

Person, M.D., Jean, *From Dog Sleds to Float Planes: Alaskan Adventures in Medicine*. Eagle River, AK: North Books, 2007.

Rearden, Jim, *Alaska: Fifty Years of Frontier Adventure*. Kenmore, WA: Epicenter Press, 2001.

Stuck, Hudson, *Ten Thousand Miles with a Dog Sled*. Lincoln, NE and London: University of Nebraska Press, 1988.

Resources at "Alaska Archives" at www.prescriptionforadventure.com

"The Mukluk Telegraph," U.S. Public Health Service, Alaska Native Health Services, Area Office, Box 7 — 741, Anchorage, Alaska.

Issues:

July 1958
August 1958
September 1958
November 1958
February 1959
April 1959
May 1959

"The Northern Lights," Tanana Day School newspaper
Charles Wheeler, editor
October 1957

John Hawkins, editor
October 1958
November 1958
December 1958
January 1959
February 1959
March 1959
May 1959

"Tanana Council News," January 17, 1959

ACKNOWLEDGMENTS

UNDERGIRDING MY WRITING about the Gaede family are my siblings and in-law siblings. Perhaps in an aviation family, it would be appropriate to say they are the wind beneath my wings: Ruth and Roger Rupp, Mark and Patti (Kvalvik) Gaede, and Mishal Gaede. Mark and Roger ungrudgingly and patiently answered my never-ending questions about airplanes, landing gears, weather, airstrips, propellers, interior configurations, and more.

Harold Gaede and Lillian Gaede Pauls, my father's siblings, made me laugh, roll my eyes, and even feel sad at anecdotes of my father's early years. In spite of their poverty, the hardworking children had energy for mischief.

Cousin Don Gaede straightened out medical jargon and procedures.

Sue Erickson Saurber has an enviable memory and provided me with conversations and descriptions about the trip to Point Hope that otherwise would have been lost.

After listening to Barbara Purbaugh Crapuchette describe her experiences of having four babies in Alaska, I was compelled to write the chapter, "Baby Can't Wait." She truly is an example of a rugged and resourceful Alaskan woman and mama.

I treasured my conversations with Manley Hot Springs school teacher, Gladys Dart, about ol' Charlie and the Innkeeper, as well as her own Last Frontier stories.

Dr. Jean Persons filled in gaps of Tanana medical history.

I couldn't have done the necessary research without the librarians at the Alaska Native Medical Center, Consortium Library at the University of Alaska-Anchorage and in Parker, Colorado. The researchers with the Project Jukebox: Oral History Program at the University of Alaska-Fairbanks were immensely gracious and accommodating, as well.

And then, there was my father, who not only stored up his stories in his head and regaled them at the supper table, but took the time to scratch them onto medical stationary in his physician handwriting. Fortunately, I managed to decipher this language before it faded away.

READER'S GUIDE

1. What comes to you mind when you think of "adventure"?
   a. Extreme sports?
   b. Challenging yourself?
   c. Doing something outside of your comfort zone?
   d. The unknown?
   e. A remote location?
   f. Danger?
   g. Thrills?
   h. Other?

2. Is your life an adventure? Yes? Describe. No? Why not? If so, of these three kinds of adventurous people, which one are you? Explain.
   a. Actively seeking to live on the edge.
   b. Living in an environment conducive to adventures.
   c. An ordinary person using his/her natural abilities and individual interests with the results translating into an adventure.

3. Why were you interested in reading this book?

4. Before reading this book, what came to mind when you envisioned Alaska? How were those perceptions changed or verified?

5. Every year, nearly one in four schoolteachers leave their jobs in Alaska, which is much higher than the national average, and Alaska leads the nation in nursing shortages. What are factors you think affect their decisions? What could help retain individuals in these positions?

6. Oftentimes, men are more attracted to a living in Alaska than women. Why do you think that is?

7. Doc and Ruby chose to go to Alaska to serve the Native people and to pay off medical school loans. What aspects, other than service to others and finances, influence a person's choice of career? What influenced yours? How has that turned out?

8. (Chp. 8) "A Strange Village Welcome") Doc was caught off guard by the murder that happened upon his arrival in Tanana. Have you been in a dispute where you feared for your safety? What did you do? Did it work? How might you have managed the situation differently?

9. (Chp.17) This chapter brings up "cabin fever." How would you have coped with the long winters of minimal daylight and severely cold temperatures? How might today's technology and social media have made a difference?

10. What were some of the challenges Doc faced as the sole physician at the Tanana hospital and along much of the Yukon River? How would you have handled such challenges? What was his attitude towards the nurses? How is that the same/not the same in hospitals elsewhere?

11. What were Doc and Ruby's attitude toward the Native people? Give examples to support your response.

12. (Chp. 26 "Flight by Faith") What is "blind faith"? Have you ever experienced it? Would you advise someone to do something in "blind faith?" Discuss.

13. Elmer and Ruby were farm kids transplanted to Alaska. What qualities did they bring with them that enabled them to be successful homesteaders?

14. Doc was larger than life, lived on the edge, always had a story to tell, and seemed unfazed by any anxiety he might be causing his wife, Ruby. Take her perspective. What might she tell you about living with this man?

15. Think about Doc and Ruby's children. By watching their parents, what were they learning about life, work, relationships, being a man, and being a woman?

16. Of the following characters, which one would you like to have known more about? Ruby? Anna Bortel? The Gronning family? Paul Carlson? Naomi, Ruth, Mark, or Mishal Gaede? Other? What is your interest in that person?

17. If you would have written this book, what would you have emphasized? Omitted?

18. Did you prefer the chapters on hunting or on flying or on medicine? Which story intrigued you the most? Discuss.

19. After reading these stories, and if it were possible, would you fly with Doc? What would persuade you one way or the other? Discuss.

20. What is *your* prescription for adventure? Discuss as a group.

MORE PRESCRIPTIONS FOR ADVENTURE

The Bush Doctor's Wife

What happens when her husband climbs off his tractor, goes to medical school, and becomes a bush pilot doctor in the middle of Alaska? She makes a home, cranks homemade ice cream on the frozen Yukon River, sings Christmas carols at 40° below zero, serves moose roasts, and seeks beadwork tips from the Native women. Follow her daily life with its frustrations and amusements.

From Kansas Wheat Fields to Alaska Tundra

Watch five-year-old Naomi grow up with village potlatches, school in a Quonset hut, the fragrance of wood stove smoke, a death-defying bush pilot doctor for a father, and a mother known for grit, hospitality, and resourcefulness. Optimism, humor, and ingenuity characterize this family as they tackle life on a homestead.

'A' is for Alaska: Teacher to the Territory

In 1954, Anna Bortel, a single woman with a teaching certificate, felt called to Valdez, Alaska, where snowfall is measured in feet, not inches. She made a snowman on top of her house. Snow sifted onto her bed. She organized the first Valdez Easter Egg Hunt — in knee-deep snow. Yet, she was undaunted, and in 1957, she headed farther north, to Tanana, where she met the Gaede family.

'A' is for Anaktuvuk: Teacher to the Nunamiut Eskimos

A trip to Anaktuvuk Pass with Dr. Elmer Gaede tugged at Anna Bortel's heart. The elders of the village had begged her to return as a permanent schoolteacher. As documented in Alaska State Legislature, Anna Bortel was the only person qualified — and willing — to accept the challenge. In this most remote village, she taught children and adults to read and write, started a craft-selling economy that enabled the people to buy food when caribou weren't present. This book gives the indigenous people a voice, records their history, and portrays the love they had for this indomitable, compassionate woman.

INDEX

C

D

E

F

G

H

I

J

K

L

M

N

O

P

R

S

T

U

V

W